D0152210

IMAGES OF ASIA

The Birds of Singapore

Titles in the series

Balinese Paintings (2nd edn.)
A. A. M. DJELANTIK

Bamboo and Rattan: Traditional Uses and Beliefs
JACQUELINE M. PIPER

Betel Chewing Traditions in South-East Asia
DAWN F. ROONEY

The Birds of Java and Bali
DEREK HOLMES and STEPHEN NASH

The Birds of Singapore
CLIVE BRIFFETT and SUTARI BIN SUPARI

The Birds of Sumatra and Kalimantan
DEREK HOLMES and STEPHEN NASH

Borobudur (2nd edn.)
JACQUES DUMARÇAY

Burmese Puppets
NOEL F. SINGER

Early Maps of South-East Asia (2nd edn.)
R. T. FELL

Folk Pottery in South-East Asia
DAWN F. ROONEY

Fruits of South-East Asia: Facts and Folklore
JACQUELINE M. PIPER

A Garden of Eden: Plant Life in South-East Asia
WENDY VEEVERS-CARTER

Gardens and Parks of Singapore
VÉRONIQUE SANSON

The House in South-East Asia
JACQUES DUMARÇAY

Images of the Buddha in Thailand
DOROTHY H. FICKLE

Indonesian Batik: Processes, Patterns and Places
SYLVIA FRASER-LU

Javanese Gamelan (2nd edn.)
JENNIFER LINDSAY

Javanese Shadow Puppets
WARD KEELER

The Kris: Mystic Weapon of the Malay World (2nd edn.)
EDWARD FREY

Life in the Javanese Kraton
AART VAN BEEK

Mammals of South-East Asia (2nd edn.)
EARL OF CRANBROOK

Musical Instruments of South-East Asia
ERIC TAYLOR

Old Bangkok
MICHAEL SMITHIES

Old Jakarta
MAYA JAYAPAL

Old Malacca
SARNIA HAYES HOYT

Old Manila
RAMÓN MA. ZARAGOZA

Old Penang
SARNIA HAYES HOYT

Old Singapore
MAYA JAYAPAL

Sarawak Crafts: Methods, Materials, and Motifs
HEIDI MUNAN

Silverware of South-East Asia
SYLVIA FRASER-LU

Songbirds in Singapore: The Growth of a Pastime
LESLEY LAYTON

White-bellied Woodpecker

The Birds of Singapore

CLIVE BRIFFETT

Illustrated by
SUTARI BIN SUPARI

KUALA LUMPUR
OXFORD UNIVERSITY PRESS
OXFORD SINGAPORE NEW YORK
1993

Oxford University Press

Oxford New York Toronto
Delhi Bombay Calcutta Madras Karachi
Kuala Lumpur Singapore Hong Kong Tokyo
Nairobi Dar es Salaam Cape Town
Melbourne Auckland Madrid

and associated companies in
Berlin Ibadan

Oxford is a trade mark of Oxford University Press

Published in the United States
by Oxford University Press, New York

First published 1993

British Library Cataloguing in Publication Data
Data available

Library of Congress Cataloging-in-Publication Data

Briffett, Clive, date.
The birds of Singapore/Clive Briffett; illustrated by
Sutari bin Supari.
p. cm.—(Images of Asia)
Includes bibliographical references and index.
ISBN 0-19-588606-2:
1. Birds—Singapore. I. Title. II. Series.
QL691.S55B75 1993
598.295957—dc20
92-41708
CIP

Typeset by Typeset Gallery Sdn. Bhd., Malaysia
Printed by Kim Hup Lee Printing Co. Pte. Ltd., Singapore
Published by Oxford University Press,
19-25, Jalan Kuchai Lama, 58200 Kuala Lumpur, Malaysia

To my children,
Anna, Peter, and Philip

Foreword

CLEAN and green Singapore: an island state strategically positioned on the Pacific Rim; a vital launching-pad into the twenty-first century; a great place for business and for holidays alike. Now, thanks to this book, you can enjoy another, more secret side of Singapore.

Lying on the southern tip of the Malay Archipelago, the islands of Singapore, large and small, are of great importance not only to the people but also to birds. The 3 million people who live their high tech lifestyle share their habitat with 154 species of resident bird, and long before Changi Airport was opened the flyways which converge on that busy place were crowded with numerous species of migratory bird, each making a short stop-over to refuel, duty free.

The fact that many of the residents and migrants can still find free green space amongst the high-rise buildings and manicured lawns of the city-state was at first by accident. They had access to what was left: scraps of mud-flats and mangroves essential for feeding, protection, and breeding, informal grasslands, and secondary and even primary rain forest around the reservoirs which keep the people and the industry supplied with the water they need.

Under the ever watchful eyes of the members of the Nature Society Singapore, these areas have been surveyed as part of a Master Plan for the Conservation of Nature presented to, and well received by, the Government. Exciting work is now under way to conserve, protect, and develop that potential. This is good news indeed, and this book will tell you why this is necessary.

Some 120 species of bird are described and illustrated in superb colour, and 64 more are discussed. So good are the descriptions and identification notes that no visitor will have an excuse to pass a bird without being able to greet it by name, or even—with practice—in its own language. Did you know that *Halcycon smyrnensis* issues a long whinnying call? Well you do now! The 'whinnyer' is otherwise known as the White-throated Kingfisher, just one of the species of kingfisher you can see lined up in sentry style along the perimeter fences of Changi Airport. Fantastic!

Most of the nations of the world, including Singapore, have signed the World Summit Directive on Biodiversity. As we move into the sustainable economy of the twenty-first century, I know that Singapore has a key role to play. The birds, so well reviewed in this little book, are the finger on the pulse of environmental care and concern. Read and conserve with care.

The Conservation Foundation, London DAVID J. BELLAMY

Acknowledgements

SINCE arriving in Singapore nine years ago, I have had the pleasure of interacting and working with most of the keen birders and conservationists in the Malayan Nature Society (Singapore Branch), now renamed the Nature Society Singapore. My knowledge of local avifauna is by no means complete or indeed equal to that acquired by some others, and I have much to thank those who have checked the text for technical corrections and competency. I am particularly indebted to R. Subharaj who meticulously reviewed the contents, and Chris Hails and Lim Kim Seng who have conducted much research and provided valuable data over the years.

Many others with whom my illustrator, Sutari bin Supari, and I have spent some fascinating bird-watching trips and have worked with on many conservation reports also deserve a mention. They include David Bradford, Richard Hale, Ho Hua Chew, Lim Kim Chuah, Lim Kim Keang, Kelvin Lim, Elizabeth Loh, Ng Bee Choo, Dick Ollington, Jonathan Smith, Morten Strange, See Swee Leng, Wee Yeow Chin, and Sunny Yeo.

I should particularly like to thank Sutari whose excellent artistic skills have not only made this book possible but highly presentable. His cheerful disposition and friendly chatter are as important to experience in the field as the birdlife itself!

Finally, my most grateful thanks go to Koh Swee Tian for her patience in typing the manuscript, to Nancy Penrose for her excellent editing contributions, and last but not least, to Hilary, my wife, in coping with the frustrations of writers and for suffering the lonely evenings and weekends.

Singapore
November 1992

CLIVE BRIFFETT

Contents

Introduction

SINGAPORE, a large, densely populated island surrounded by some thirty small offshore islands, is strategically located at the southern tip of the Malay Peninsula in South-East Asia. Visitors are impressed with its clean and green urban environment but are surprised to find that natural habitats such as primary and secondary forests, mangroves, freshwater reservoirs, and associated marsh areas are still present in various parts of this island state. Avian tourists also recognize the value of existing habitats and ecosystems and readily use them as a stop off point to feed and rest twice a year during annual migration movements from northern China and Siberia to Australasia. These migrant birds range from the larger prey species comprising eagles, harriers, kites, and hawks to smaller passarines such as flycatchers, warblers, and shrikes. The longer-distance wader migrants, ranging from the larger godwits and whimbrels to the very small plovers, sandpipers, and stints, all require extensive mud-flats on which to feed. Singapore also attracts a number of the world's rare bird species, including the migrating Chinese Egret, Asian Dowitcher, and Nordmann's Greenshank. The resident species, present for most of the year, occupy numerous habitats and range from the largest bird of prey, the Grey-headed Fish-eagle, to the smallest warbler, the Flyeater.

This book describes 120 bird species, all of which are illustrated in colour, principally the male species. A further 64 birds, some of which are shown in black-and-white sketches, are referred to in the text to provide good coverage of the families present and to aid identification. Where certain species have distinctive calls, these are noted. Reference is also made to typical habitats where they are commonly found. An indication of distribution, occurrence, and abundance is also included in the section on status and in the Checklist designations.

Because of Singapore's relatively small size and the demands of its population, most bird habitats are under threat. In 1990 the Malayan

Asian Dowitcher

Nordmann's Greenshank

Nature Society of Singapore published a Master Plan for the Conservation of Nature to identify the most important areas for wildlife (see Map 1). Fortunately, this was well received by government ministries. Numerous planning proposals and environmental projects are now under way to try and conserve the best of these areas, but obviously some form of compromise is required to meet the continuing built development needs of the population. There are presently several 'protected' areas, including the centrally located Bukit Timah and Central Catchment Nature Reserves, that accommodate most of the forest bird species. Since these reserves are much reduced from their original areas and have been subjected to numerous disturbances, many forest bird species have become extinct. If, however, no further encroachments or major changes in habitat provision take place, the present resident and visiting population of over 190 recorded species should continue to use this important and rich natural resource and hopefully their abundance will also increase.

Grey-headed Fish-eagle

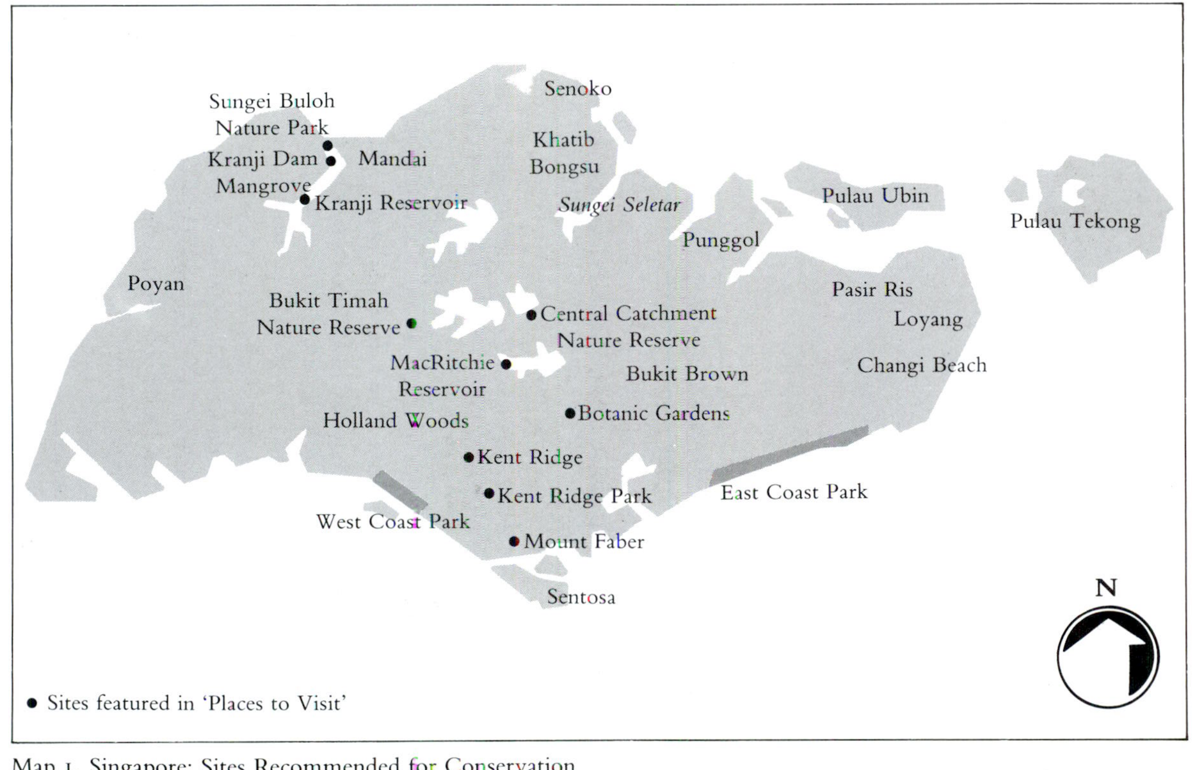

Map 1. Singapore: Sites Recommended for Conservation

Mangrove areas have been severely reduced along the coastlines over the years and the best of those remaining are to be found at Mandai, Sungei Buloh, Kranji, Sungei Seletar, and Pasir Ris on the mainland and on Pulau Ubin and Pulau Tekong, the two largest northern off-shore islands. A number of resident bird species totally dependent on this habitat can still be found in these mangroves. The wild bird nature park created at Sungei Buloh includes mangrove habitat and is the first of its kind for Singapore. Special visitor facilities are provided, including an educational centre, mangrove boardwalk, nature trails, viewing hides, and an arboretum area for mangrove propagation and research. Details of this reserve and other interesting bird-watching areas are covered in 'Places to Visit'.

The numerous reservoirs of Singapore are also a further potential source of bird habitats but only a few have adjoining marsh areas that are particularly rich in bird and other animal life. Such places are Kranji Reservoir, where a Grey Heronry has become well established in recent years, and Poyan on the west coast, where Pheasant-tailed Jacanas and other waterfowl rare to Singapore, such as the Common Coot, Cotton Pygmy Goose, and Garganey, are seen. Freshwater marsh areas also harbour many crakes, rails, and the Purple Swamphen. They provide nesting habitats for Yellow and Cinnamon Bittern, and attract large flocks of White-winged Tern. Undoubtedly, the richest area for freshwater life is the swamp forest in the Central Catchment at Nee Soon. Here can be found some fascinating mammals, such as the Banded Leaf Monkey and Cream-coloured Giant Squirrel, both endemic sub-species to the island, as well as many species of freshwater and other endangered invertebrate, including crabs, snakes, and frogs. Vulnerable birdlife found in this area includes the White-chested Babbler, Moustached Babbler, and Blue-eared Kingfisher.

Grassland habitats also deserve a mention as they attract certain bird species, such as weavers, munias, and prinias, and give excellent cover for other birds and animals. Such areas often develop on reclaimed land and wasteland; where water pools and streams are present, a good variety of resident and migratory birdlife can be found. Recent proposals to reduce grass-cutting maintenance and create long-grass bird sanctuaries within parks will certainly assist in increasing the diversity and abundance of such birds, provided the areas are large enough. More

White-chested Babbler

Common Coot

Pheasant-tailed Jacana

opportunities are also available on the many golf courses by creating 'rough' and by establishing long-reed and long-grass areas around ponds to encourage birdlife habitation.

Other initiatives being pursued by government departments include the formation of an islandwide green corridor network, a redesign of existing water body areas along river culverts and coastal areas, and the conversion of parks and gardens to increase plant diversity suited to attract more wildlife. These proposals are exciting and will be beneficial to the survival of wild birds. It is hoped that this book will serve to encourage more people to watch, identify, and enjoy the birds of Singapore, whose future occurrence and abundance will depend on both the support and influence of the local peoples in protecting the natural habitats and ecosystems in which these birds survive.

Bird Status and Distribution

In the last ten years the substantial amount of bird-watching that has taken place in Singapore has added to our knowledge of the occurrence and status of wild birds. Such activity, if combined with careful note-taking, and if within the ambit of structured and well-organized surveys, censuses, and races, can contribute to building up a bank of research data of considerable value for conservation management. Such observations should not be confined merely to recording bird sightings but should also include an assessment of habitat provision, evaluations of the impact of various built developments, the monitoring of pollution problems, and the identification of other threats such as catching and hunting activities.

At the present time, the Singapore Checklist included at the end of this book comprises 326 species, of which 118 are confirmed resident breeding birds. A further 36 species are thought to be breeding but many of these have very low populations. The total number of species has increased by 42 since the first modern-day checklist was collated in 1984, but many of these are sightings of vagrants or rare migrants and their addition is probably due more to an increased number of bird-watchers than any major improvements in habitat provision!

Since records began, it has been estimated that at least 87 species of forest bird have already been lost; this represents over 80 per cent of those birds which have become extinct in Singapore. Further depletion and increased disturbances to the remaining forest areas will critically affect those species left. Recent research confirms that 52 species of bird occurring in Singapore today are believed to be at risk of extinction. Of these, 37 are found in the primary and secondary forest areas and 12 inhabit mangroves.

Since an estimated 94 species occur on regular annual migration, and at least 40 of these are totally dependent on mud-flats, the conservation of this form of habitat is also particularly important for the birds. Unfortunately, numerous additional reclamation schemes for almost all the southern islands and at Changi, Pasir Ris, Serangoon, Mandai, Tekong, and Ubin are going to substantially reduce mud-flats in the future. Such proposals also threaten the survival of some resident species, including the Beach Thick-knee, Great-billed Heron, and Pacific Reef-egret, which are already amongst the rarest birds of Singapore.

The Checklist designates the status of each species as either a resident or a visitor. Visitors may be passage migrants or winter visitors. The migrants usually stay for a few days, or even weeks, *en route* to their breeding grounds to the north or winter quarters further south. (Winter and summer designations used in this book refer to the northern temperate zone seasons for ease of reference.) Winter visitors usually stay for longer periods than passage migrants, often extending their

Beach Thick-knee

sojourn for several months between September and March. In many cases, some winter visitors may join a local population of resident birds or also be defined as passage migrants.

A non-breeding visitor in this book is defined as a species that is not known to undertake a definite migration journey. It could occur at any time of the year and may well breed in adjoining countries, such as Indonesia or Malaysia.

There is only one species—the Blue-throated Bee-eater—given the status of migrant breeder. It arrives in February or March and leaves on its southward journey in September. Additional categories include introduced birds, which are often aggressive and attain higher pest populations, and escapees, which need to have proved a self-sustaining facility in the wild to be included in the Checklist. Readers may also find helpful, although somewhat subjective in accuracy, the abundance guidelines that indicate how easy or difficult it is to see a particular species and whether or not it is restricted to particular locations.

It is likely that further modifications to status and distribution will be made in the future as more information is obtained through observation and ringing techniques.

Places to Visit

THE full value of a bird guide depends not only on providing illustrated species to aid identification but also on defining the places where the birds may be found. This chapter is therefore devoted to highlighting the principal places most frequented by birds and bird-watchers in Singapore. An attempt has been made in this selection to cover a wide range of habitats and to choose sites not fully documented in previously published texts. Where repetition does occur, the information provided here represents the latest situation.

Primary Forest: Bukit Timah Nature Reserve

The unique character of this site, once described by David Bellamy as the 'richest ecological high-rise estate in Singapore with a natural air-conditioning system that costs nothing', warrants its inclusion. As fuller details are obtainable elsewhere on other flora and fauna, it is simply intended here to confine comments to bird-watching activities.

The very presence of a primary forest in Singapore is in itself surprising. Bukit Timah's reputation of being the most accessible and perhaps the oldest rain forest reserve in the world warrants continued conservation for the benefit of human visitors, but mostly for the regeneration of its rich ecological resources. Despite its relatively small size of 81 hectares, and its island status which was acquired in recent years by the construction of a new expressway, and by other threats caused by former quarry workings and the use of the reserve for telecommunication and military purposes, Bukit Timah still provides many bird-watching surprises (see Map 2). At the lowest entrance point to this 162.5-metre-high granite hill, an interpretative centre, which has a car park, is provided to educate visitors. Here the visitor can find literature to guide his expedition, which may comprise a direct, steep, arduous walk to the summit along a tarmac road or a more leisurely natural trail along the contour paths. Familiarization with flora and

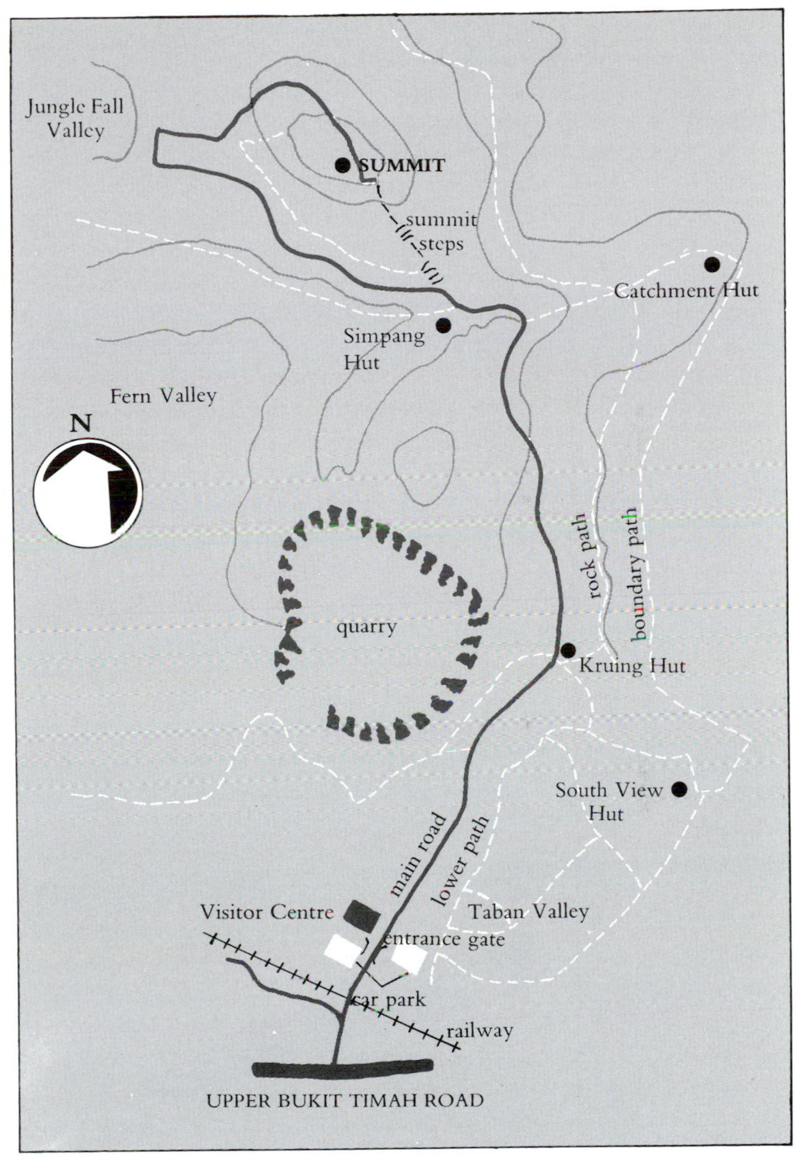

Map 2. Bukit Timah Nature Reserve

fauna in this reserve is recommended and proceeding with patience and quietness can be rewarding. Using caution and sharp observation skills, it may be possible to see a number of rare bird species on the forest floor during the migrant seasons. These include the White-throated Rock-thrush, Orange-headed Thrush, Hooded Pitta, and Siberian Blue Robin. In the lower shrubs and ferns several species, including the Short-tailed Babbler and Chestnut-winged Babbler, are more likely to be heard than seen, along with the Little Spiderhunter and the very common Striped Tit-babbler. At mid-canopy level common residents include the spectacular and bizarre-shaped Greater Racket-tailed Drongo, which is famous for its mimicry skills of other birds, and the beautiful Asian Fairy-bluebird. Colourful and noisy species seen feeding on tree-trunks and using nesting holes include the Banded Woodpecker and the Red-crowned Barbet and sometimes the rarer Blue-rumped Parrot. In the top canopy, which is more easily viewed from upper levels of the summit, the Blue-winged and Lesser Green Leafbirds may first be heard and then seen by the lucky bird-watcher, along with the Scarlet-backed Flowerpecker and several sunbird species. In the tops of dead trees the Dollarbird, Brahminy Kite, and White-bellied Sea-eagle, all of which nest in the reserve, may be observed.

Reaching the summit is worth the effort as it affords a magnificent view of the Central Catchment Nature Reserve and provides more opportunities to study aerial feeding birds. These include the resident Edible-nest and Black-nest Swiflets (too difficult to distinguish in flight), the House Swift with its white rump, the Pacific Swallow and

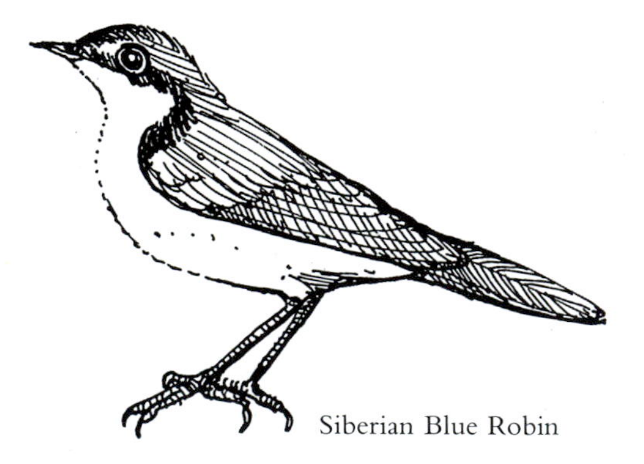

Siberian Blue Robin

Barn Swallow, and the Grey-rumped Treeswift, which has long, pointed tail streamers. During periods of migration, the larger Fork-tailed Swift (also with a white rump), and the remarkably fast-flying White-vented Needletail regularly occur.

Bukit Timah can be frustrating, exasperating, and disappointing for bird-watching, especially at midday. The early morning or late afternoon are the best times. Visitors need to be patient and observant, and be prepared to walk some of the contour paths. On a lucky day, the rewards can be better than at any other location in Singapore, for Bukit Timah is exceedingly diverse in species, unique in habitat, and spectacular in appearance.

Blue-rumped Parrot

White-vented Needletail

Secondary Forest: Central Catchment Nature Reserve

Located in the centre of the main island of Singapore, this important reserve contains many interesting and productive ecosystems which attract at least 190 species of resident and migrant bird. It is the largest area of protected reserve on the island and as such provides the greatest contribution to the maintenance and self-sustainability of birdlife.

However, its further degradation by built developments, golf courses, and industrial plants, and its overuse for military purposes has unquestionably resulted in a severe reduction of the natural fauna in Singapore.

The reserve comprises four reservoirs, a well-established secondary forest that has developed over the last eighty years from former abandoned plantations, and remnants of some primary tree species located around the MacRitchie Reservoir. There are many entry points but the easiest way in, and perhaps the most rewarding for seeing birds, is along Sime Road through the Singapore Island Country Club (see Map 3). By keeping to the tracks shown, it is possible to see many forest species. A telescope is recommended for obtaining distant views across the golf courses, over marshy areas, or along forest streams

In this reserve at least six species of kingfisher may be seen during the migration period. The Blue-eared Kingfisher is probably the most difficult to sight, being an extremely rare resident that chooses the thick, forested streams as its habitat. The other local residents, including the Collared, White-throated, and Stork-billed Kingfishers, are more regularly encountered. The Black-capped and Common Kingfishers are also frequently occurring migrants.

Woodpeckers have been on the decline over the last forty years. The White-bellied Woodpecker featured on the frontispiece to this book is a former resident that has not been seen since 1988, although it may still be present in this reserve. Rufous and Banded Woodpeckers are common and the diminutive Brown-capped Woodpecker is also regular but more often occurs outside the forest in the golf course trees. Fig trees in fruit along the main tracks attract large numbers of pigeons, orioles, barbets, and bulbuls. These include the Jambu Fruit-Dove, Thick-billed Pigeon, and Red-crowned Barbet. The prolific *Simpoh Ayer* shrub bordering many of the tracks produces succulent red seeds as the buds burst open in the morning sun, and Olive-winged, Yellow-vented, Cream-vented, and Red-eyed Bulbuls usually appear for a feast. Olive-backed and Brown-throated Sunbirds also partake of the seeds, while Long-tailed Parakeets use their dextrous feet to rip off the buds before they open. Drongos and bluebirds also inhabit the secondary forest, along with a number of migrant cuckoo species and the resident

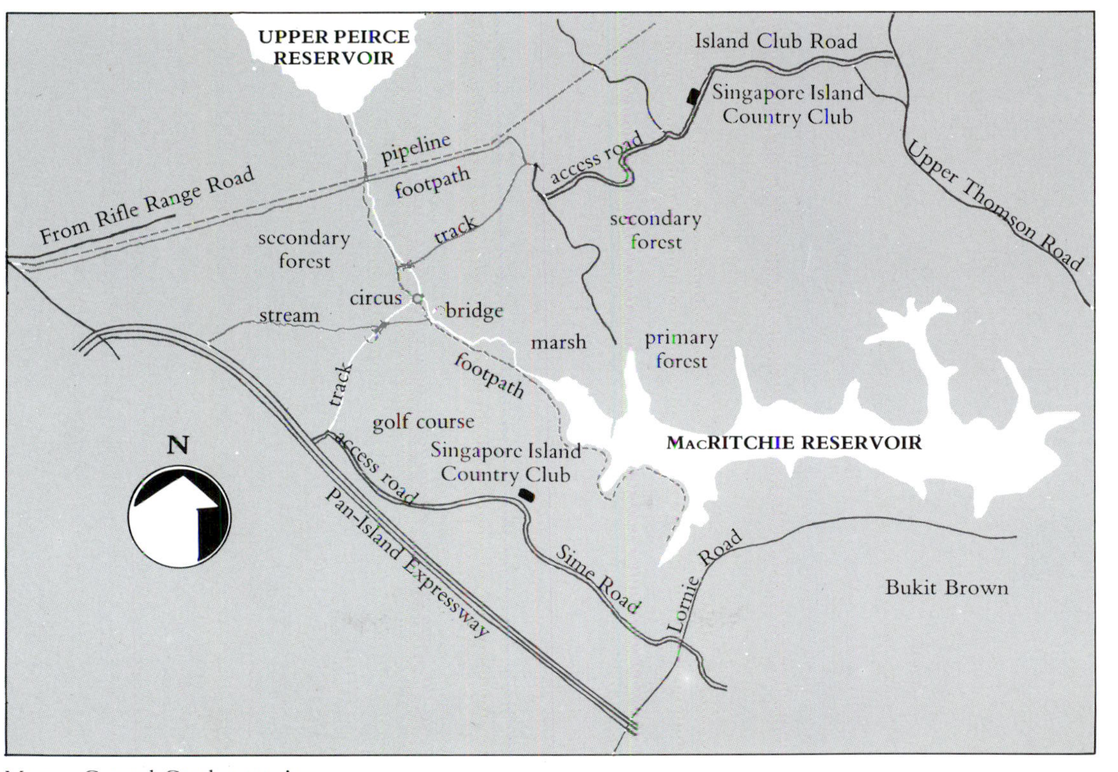

Map 3. Central Catchment Area

Jambu Fruit-Dove

Chestnut-bellied Malkoha that moves quietly through the mid-storey of the tree canopy.

Around the reservoirs waterfowl occurrence is minimal because of the excessive landscaping husbandry that has created short-grass perimeters and few water plants. Aerial feeders, such as the Blue-throated and Blue-tailed Bee-eaters and the Pacific and Barn Swallows, glide leisurely over the surface waters. Several birds of prey may be seen, including the rare Grey-headed Fish-eagle and Crested Serpent-eagle, both of which may still nest in the forest. After dark the distinctive call of the rare Malaysian Eared Nightjar is frequently heard, along with the Brown Hawk-owl and the Collared Scops-owl, two other common residents. The special treat of hearing the Spotted Wood-owl emphasizes the importance of conserving this reserve in its entirety to protect endangered species for the benefit of future generations of humans to enjoy and for wildlife to survive.

Mangroves and Mud-flats: Sungei Buloh Nature Park

Sungei Buloh is a coastal site of 87 hectares located in the north-west corner of the main island of Singapore. It comprises 40 hectares of prawn ponds enclosed by two small rivers and a 9.5-hectare offshore island of mature mangrove (Pulau Buloh). Some inland mangroves along the rivers remained after initial clearance for the bunded pond system and all these areas are accessible from the sea by sluice gates. To the west side are 6 hectares of freshwater ponds served by drainage water from the hinterland now being developed for agricultural-technology use. Other backland areas free of tidal undulation are enclosed by a security fence and comprise coconut and other fruit tree plantations.

After two years of development, Singapore's first wild bird nature park is almost ready for public use. A considerable amount of earth-moving and change to existing bunds has been carried out to improve facilities for migrant wading birds and to enable visitors to view these without causing disturbance. Several access trails extending outwards from the new visitor centre that is located to the east side of the Sungei Buloh Besar lead to numerous observation hides (see Map 4).

After visiting the centre, which depicts the flora, fauna, and mangrove ecology systems found on this reserve, a walk is recommended along the boardwalk through the mangroves. Here, one is most likely to see several species of heron and kingfisher and observe Brown-capped and Rufous Woodpeckers. The Pied Fantail, which is an excitable and acrobatic black-and-white bird, the Ashy Tailorbird, and several sunbird species, including the Copper-throated Sunbird, may also be observed. The Mangrove Whistler, a species indigenous to mangrove habitat, sometimes occurs, along with more common species such as the Flyeater, Common Iora, and Pied Triller, all of which have also managed to adapt to other habitats in Singapore.

At the largest hide across the river bridge, is the first opportunity to view waders that feed on the mud-flats during the migrating season. The most common waders seen here include the Pacific Golden Plover, Common Redshank, and Marsh Sandpiper, and the Rufous-necked Stint and Whimbrel, with their long, decurved bills. Rarer visitors can be picked out by careful observation through the telescope and these may include the Asian Dowitcher, Broad-billed Sandpiper, and Great Knot.

Along the trails on the inland side, many bird species occur in the secondary shrub and plantations, including the Olive-winged and Yellow-vented Bulbuls, Dollarbird, Black-naped Oriole, Spotted Dove,

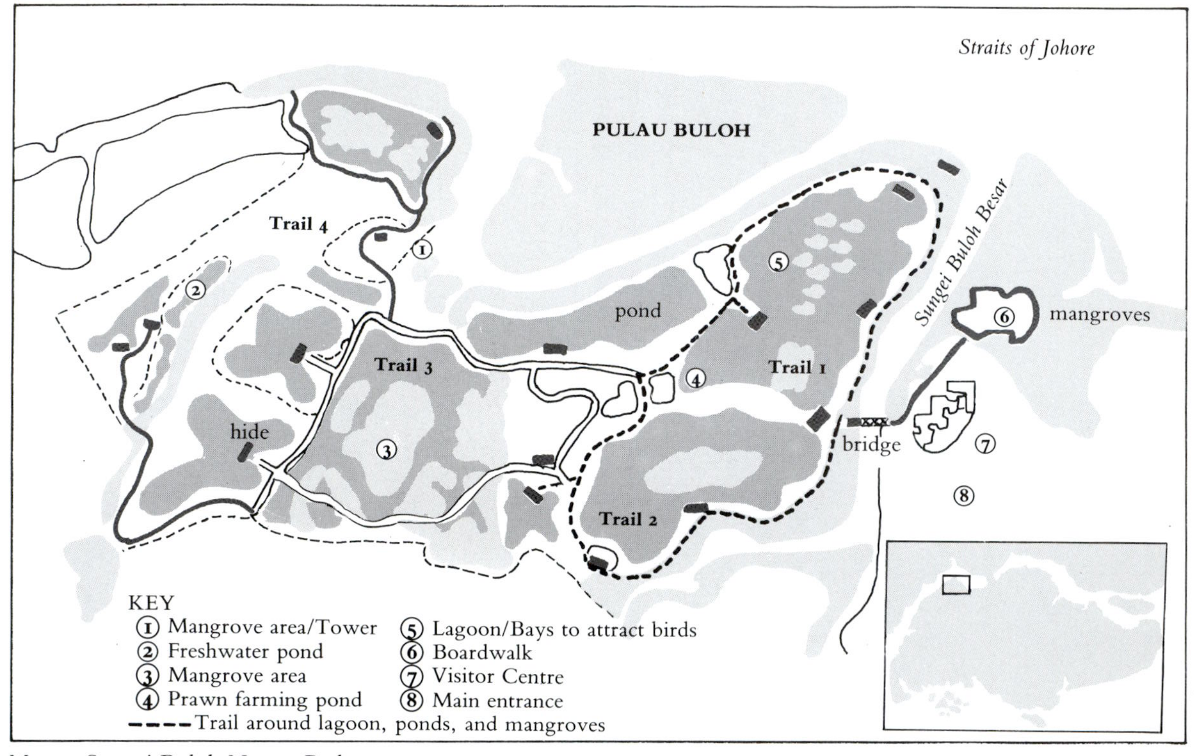
Straits of Johore
PULAU BULOH
Sungei Buloh Besar
mangroves
Trail 4
Trail 3
Trail 1
Trail 2
pond
hide
bridge
KEY
1 Mangrove area/Tower
2 Freshwater pond
3 Mangrove area
4 Prawn farming pond
5 Lagoon/Bays to attract birds
6 Boardwalk
7 Visitor Centre
8 Main entrance
Trail around lagoon, ponds, and mangroves

Map 4. Sungei Buloh Nature Park

Great Knot

Peaceful Dove, and also the perky Magpie Robin. Migrant passarines include several species of cuckoo, warbler, flycatcher, and shrike. On the outer bund, views of terns are possible and large flocks of waders or egrets may be seen wheeling over the sea returning from offshore mud-flats to the reserve as the tide levels rise. Over-flying birds of prey, such as Ospreys and White-bellied Sea-eagles, are commonly seen fishing the waters, with Brahminy Kites feeding on floating scraps. At the freshwater ponds, further delights await the keen birder with the presence of herons and bitterns, the Lesser Treeduck, and the Common Moorhen, and, possibly, the Garganey.

The control of water levels on this reserve is critical for attracting the greatest number of birds. A management plan prepared by overseas consultants and the Jurong Bird Park establishes the arrangements needed.

Other areas of significance comprising mangrove and mud-flat habitats include Senoko (Ulu Sembawang), Pasir Ris, Sungei Mandai, Kranji Dam Mangrove, and Sungei Seletar, all of which are located along the more sheltered northern shores of the main island of Singapore (see Map 1). Pulau Ubin, an offshore island, can be reached by a 15-minute boat trip from Changi and has more mangrove areas to investigate. Pulau Tekong, a larger island, is currently a military area but hopefully will, in the near future, be reopened for public access and use.

Freshwater Marshes: Kranji Reservoir

Situated close to Sungei Buloh and formerly comprising a mangrove reserve, the Kranji Reservoir today is a freshwater habitat. The most important areas for birds are the edges that still have sufficient marsh areas to provide the needs of many species. The long grass adjoining the marsh is also important for such species as crakes, rails, and waterhens. Here, the visitor may see the White-browed Crake, the enormous chicken-sized Purple Swamphen, and the Yellow and Cinnamon Bittern, both of which breed in the reservoir. The Black Bittern also occurs on migration along with the Oriental Reed-warbler

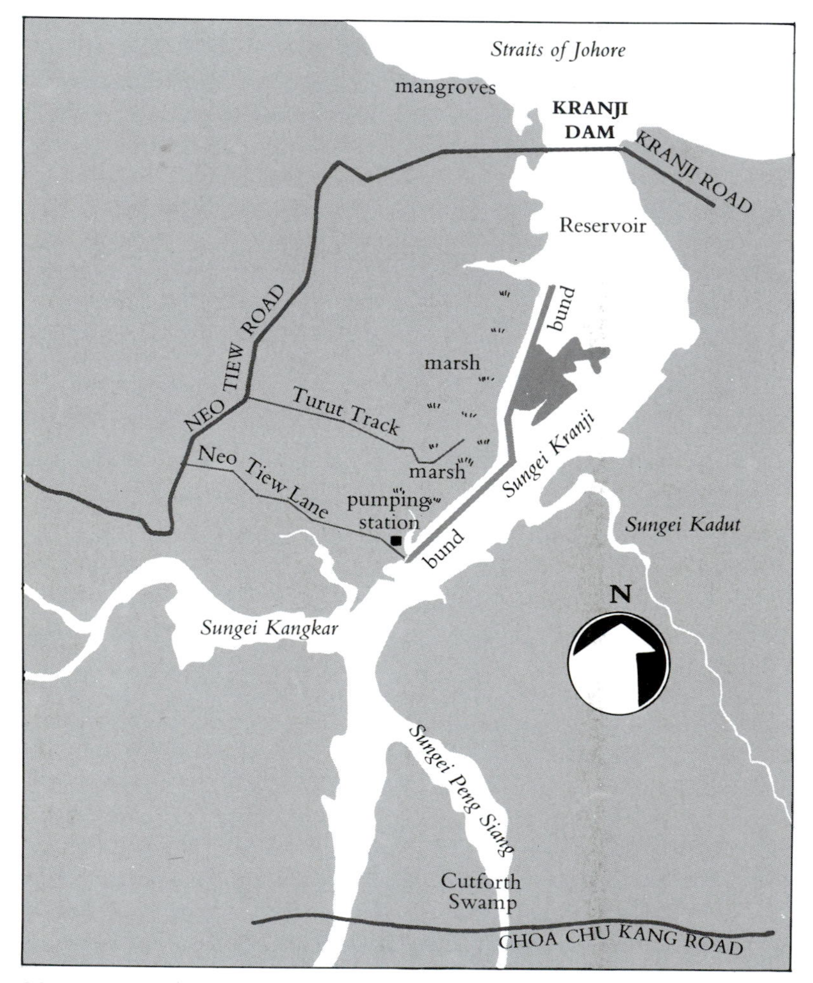

Map 5. Kranji Reservoir

and Black-browed Reed-warbler. Over the water one may see large numbers of White-winged Tern and Little Tern, and there are plenty of resident Grey and Purple Herons that nest in the high trees around the reservoir. Birds of prey include the Black-shouldered Kite, Brahminy Kite, and Osprey. The White-bellied Sea-eagle nests in the high trees close to the reservoir.

Entry points shown on Map 5 include Neo Tiew Road to the PUB pumping station to the west, Kranji Dam to the north, Sungei Kadut to the east, and Cutforth Swamp to the south.

Parks and Gardens: Botanic Gardens

This site is located close to the heart of the commercial centre of Orchard Road and has been in existence for over 130 years (see Map 6). Recently, many improvements have been made to enhance the natural features of the gardens, but being gardens of botanical interest they obviously have a certain degree of formality and are subject to constant husbandry. The wide diversity of plants, including a profusion of flowers and some remarkably impressive trees of great age, still attracts a wider range of birds than most other gardens and parks in Singapore. Remnants of a primary forest, though small, remain to fascinate the visitor, and the forest alongside Tyersall Avenue also contains some primary tree species. Walking through these forested areas in the early morning, one will be rewarded with views of the Banded Woodpecker, Hill Myna, Abbott's Babbler, and Striped Tit-babbler, along with a few migrants such as the Asian Brown Flycatcher and Arctic Warbler.

High in the trees around the gardens, Long-tailed Parakeets are usually seen and the Goffin's Cockatoo, an escaped species now self-sustaining in the wild, is always heard. A rare visitor that has attempted to breed in recent years is the Crested Goshawk. The Japanese Sparrowhawk, a common migrant, and the Brahminy Kite, a resident, are not uncommon in the gardens. Treetop species, such as the Scarlet-backed Flowerpecker, Philippine Glossy Starling, and Pink-necked Pigeon, are also common. Other species using the mid-canopy include the Common Iora, Pied Triller, and Brown-capped Woodpecker. A specialist feeder found nesting in the fan palms and difficult to find

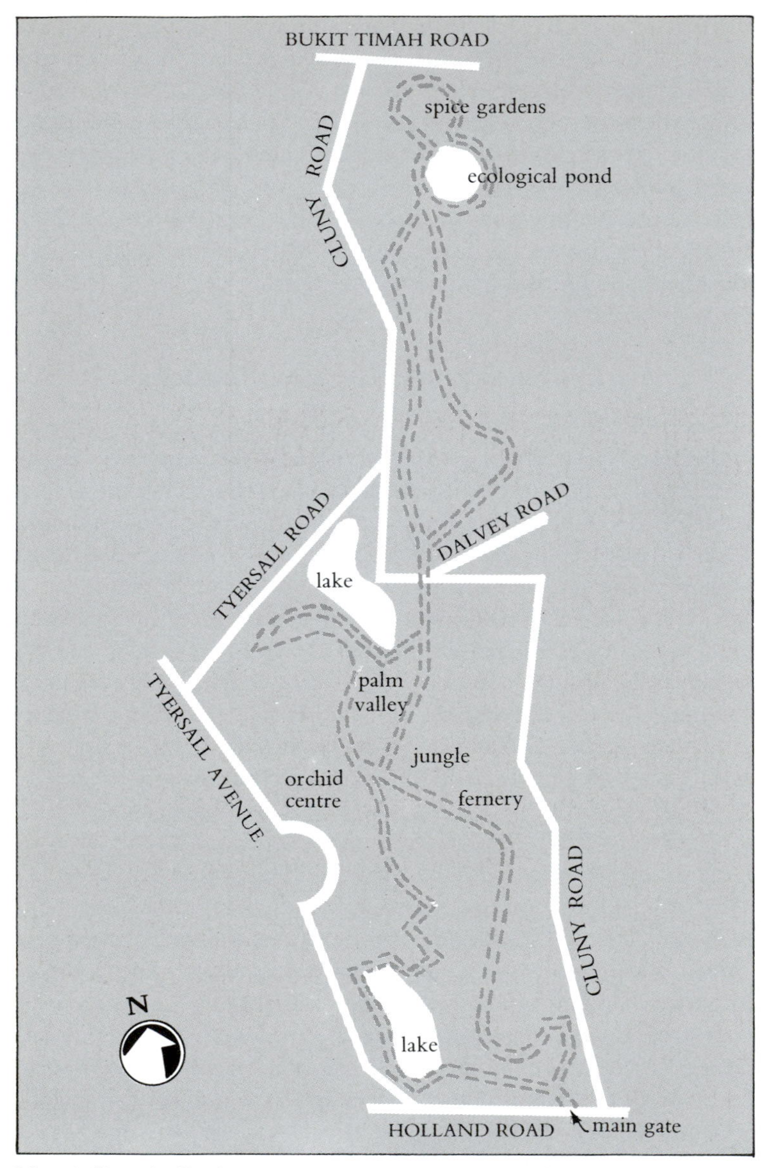

Map 6. Botanic Gardens

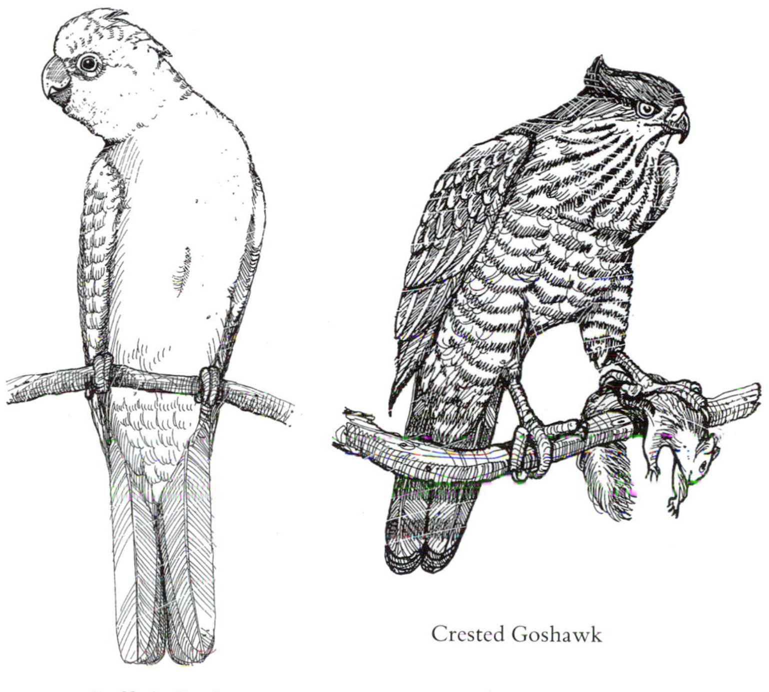

Crested Goshawk

Goffin's Cockatoo

elsewhere in Singapore, is the Asian Palmswift. Regular residents using the flowering trees and plants are the Olive-backed and Brown-throated Sunbirds. Scaly-breasted Munia can be found in the long-grass areas. On the lawns, the commonest birds following the grass cutters are the White-vented Myna and a few Magpie Robin that continue to survive thanks to some reintroductions made a few years ago. The Yellow-vented Bulbul may be seen nesting even inside 'horses' in the topiary area, and the Common Tailorbird constructs nests in delicate cups formed by 'sewing' two large leaves together.

The proposed ecological pond in the extension to the gardens will be completed in 1994 and is likely to attract an increasing number of water birds, including the Lesser Treeduck, White-breasted Waterhen, Yellow Bittern, and Common Kingfisher. Perhaps the most obvious birds seen throughout the gardens are the Collared Kingfisher, a noisy but attractive blue-and-white bird, and the Black-naped Oriole, a remarkably beautiful yellow bird with a musical, fluty call.

Other areas with parkland habitats include the West Coast Park, which has a marsh garden, and the East Coast Park, with its long-grass bird sanctuary areas. Holland Road and Bukit Brown also comprise larger garden-type habitats with good numbers of wayside trees, shrubs, and long-grass areas (see Map 1 for locations).

Secondary Woodland: Mount Faber to Kent Ridge

A recent proposal to establish connections between existing trails along several hilltops rising above the south-west coastal area of Singapore would make it possible to commence from Mount Faber and walk across Telok Blangah and Kent Ridge Park to Kent Ridge itself and then on to Clementi Woods and to the West Coast Park without having to cross a road. Each of these locations offers a range of

Crested Honey-buzzard

Common Buzzard

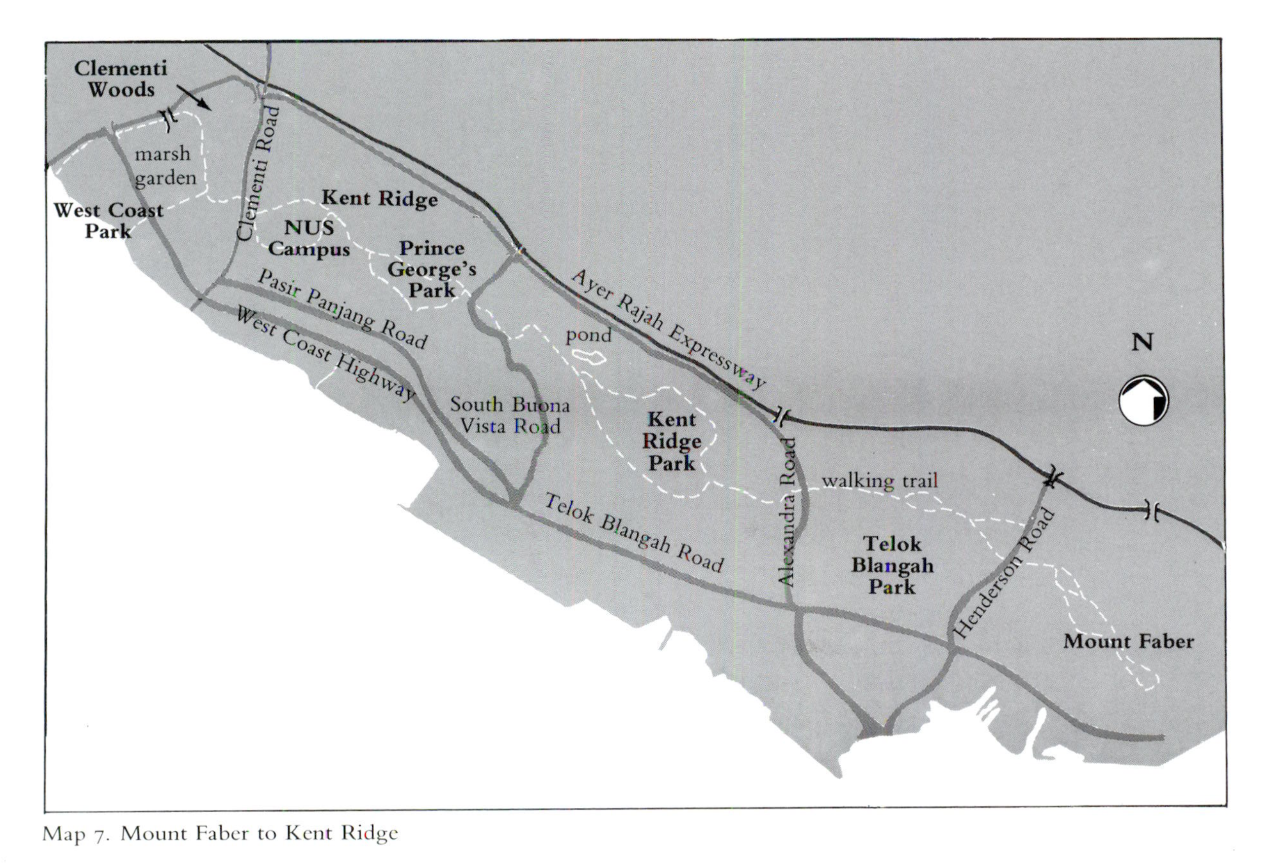

Map 7. Mount Faber to Kent Ridge

vegetation and habitat mainly comprising secondary woodland intermingled with park-style planted trees and grassy areas. There is also a freshwater pond in Kent Ridge Park. Panoramic views are available on both sides of the ridge which make this a most pleasant place for a walk or jog. For the bird-watcher, the higher elevation enables soaring migratory and resident birds of prey to be seen. These include the migratory Crested Honey-buzzard, Common Buzzard, Black Baza, and Japanese Sparrowhawk. Residents comprise the Black-shouldered Kite, which nests near Clementi Woods, and the Brahminy Kite and White-bellied Sea-eagle, both of which nest at Kent Ridge.

The usual park birds noted in the Botanic Gardens all occur here and, in addition, aerial feeders, such as the Blue-throated and Blue-tailed Bee-eaters, Dollarbird, swiftlets, needletails, swifts, and swallows, are regularly sighted.

Entry may be made at any point along the trail as shown on Map 7.

The Birds of Singapore

IN this section, 184 species of the most commonly occurring birds in Singapore are reviewed. Comments on their habits, habitats, status, distribution, and special distinguishing features are included, especially when these enable closely related species to be separated. More obvious colouring, shape, and structural characteristics may be observed in the coloured illustrations. Each bird belongs to a family comprising a number of closely related species. General characteristics and habits are provided as an introduction to each group. Where possible, additional reference is made to other closely affiliated species and, in some cases, these are illustrated with black-and-white sketches.

Herons, Egrets, and Bitterns

These commonly occur close to water where, using a long-legged wading and stealth technique, they spear fish, frogs, and other underwater life with sharp, dagger-shaped bills. They are of larger size than most birds, with long necks often retracted when in flight. The egrets are mainly white and develop attractive neck and head plumes during the breeding season. The colour of their bills, legs, and feet and their overall size are the best distinguishing features. The herons have a slow, ponderous flight and land awkwardly in long grass and reeds near the water's edge or in trees in which they breed.

GREY HERON

Ardea cinerea 102 cm (40″) Plate 1a

This impressive, large bird occurs mainly in freshwater reservoir areas and is also seen along river estuaries with shallow-water edges. It usually feeds singly on fish and frogs, but roosts and breeds communally. A large heronry has been established at Kranji Reservoir in recent years, but because of recent development works the breeding birds have been disturbed and there is an urgent need to provide alternative, suitably protected sites in the same areas. The similar but even bigger GREAT-BILLED HERON *Ardea sumatrana*, which is the largest (115 cm) bird in Singapore, is much rarer and only occurs in marine habitats on the southern islands. Because of reclamation disturbances, this species is highly endangered in Singapore and present known feeding sites off Pulau Bukum, Pulau Hantu, Sentosa, and the Jurong River are in need of urgent conservation.

Great-billed Heron

PURPLE HERON

Ardea purpurea 97 cm (38″) Plate 1b

More slender-necked and of slightly smaller size and darker plumage than the Grey Heron, the Purple Heron is found in similar habitats though it is generally more widespread. It mainly nests in trees at the Kranji Reservoir but also uses some large ground fern nesting sites in the marsh. It is a solitary, inconspicuous feeder against tall reeds, using a stalking and quick jabbing hunting technique. The purple back and wing feathers are prominent in flight, and a pronounced kink in the neck is also visible.

LITTLE HERON

Butorides striatus 46 cm (18″) Plate 1c

The Little Heron is a small, squat bird that adopts a crouching stance and prefers to feed on mud-flats in shallow water. It occurs along all coastal areas of Singapore and near inland rivers, and may even be found around ponds and lakes in parks and gardens. It usually feeds singly and breeds mainly in mangrove habitats, and can often be seen feeding along the tide-line with other wading birds. It emits a sharp 'keyow' call when disturbed. The CHINESE POND-HERON *Ardeola bacchus* is of similar size and shape but is more prone to inhabit freshwater marsh areas and only occurs in small numbers in Singapore as a passage migrant. The non-breeding plumage comprises heavy brown streaking to the breast but in flight the white wings are distinctive.

CATTLE EGRET

Bubulcus ibis 51 cm (20″) Plate 1d

A common winter visitor occurring in large flocks, the Cattle Egret is usually found inhabiting inland freshwater marshes, sewage farms, and narrow water courses. It is distinguished from other egrets by a short or stubby yellow bill, smaller overall size, and stockier shape. In May, there is evidence of its distinctive breeding plumage, comprising an orange head, throat, and upper back feathers.

GREAT EGRET

Casmerodius albus 89 cm (35″) Plate 1e

The largest of a number of entirely white-plumed birds, the Great Egret is most easily distinguished by its comparatively large size, huge yellow bill that turns black in the breeding season, and the odd kink in its neck. Its legs, including the toes, are black. The similar but smaller LITTLE EGRET *Egretta garzetta* has a smaller black bill and legs with yellow toes. This bird is seen mostly near coastal and fish pond areas and inland mud-flats. Its numbers increase during the non-breeding period, providing a spectacular fly-past with other egrets.

PACIFIC REEF-EGRET

Egretta sacra 58 cm (23″) Plate 1f

This bird occurs in two colour phases, either totally white or dark, slaty grey. It is regularly seen feeding singly from rocky outcrops or along river courses on the outlying islands. In flight the legs project beyond the tail much less than with other egrets. In its white phase, it is difficult to distinguish from the CHINESE EGRET *Egretta eulophotes* although the latter is larger. In non-breeding plumage, the Chinese Egret, a world-endangered species, has yellowish-green legs and a black bill sometimes with a yellow base to the lower mandible. Its bill length is equal to the length of the upper tarsus. It is usually recorded annually in small numbers.

Chinese Egret

BLACK-CROWNED NIGHT-HERON

Nycticorax nycticorax 61 cm (24″) Plate 2a

An attractive shorter legged heron, the Black-crowned Night-heron is most active during dawn and dusk, but it also feeds throughout the night. The adult bird has a sinister appearance when viewed closely, with its white head plumes, black cap, and blood red eyes. The immature colouring of light buff spots on a dark brown plumage provides a total contrast to the black, grey, and white plumage of the adult. The bird is particularly fond of mangrove habitats that provide suitable shaded roosting and nesting facilities. A large heronry of over 300 pairs was discovered in 1988 at Khatib Bongsu, on a mangrove island, but because of disturbances most birds have now left this location although, it is hoped, only temporarily.

YELLOW BITTERN

Ixobrychus sinensis 38 cm (15″) Plate 2b

The Yellow Bittern is a smaller family member, often inconspicuous when nesting or feeding in long grass or freshwater marsh areas. When flushed out of the reeds, it is easily distinguished by its mustard yellow-and-black wings but immature birds have dark, streaky-brown breasts. It occurs throughout the island wherever a water-filled ditch, marsh, pond, river, or reservoir is present and where there is also good ground cover. Some birds are resident breeders, but numbers are substantially enhanced by non-breeding visitors during the northern winter time. A much larger and less common winter visitor is the highly secretive BLACK BITTERN *Ixobrychus flavicollis* with its black back and wings and distinctive large, yellowish-brown neck patch. This species has been seen annually in Kranji, Sungei Buloh, and Punggol.

CINNAMON BITTERN

Ixobrychus cinnamomeus 38 cm (15″) Plate 2c

A common resident, the Cinnamon Bittern is less numerous than the Yellow Bittern though it occupies similar habitats. Adult birds are easily recognized in flight by their rich chestnut colouring on the back and wings. The immature birds are heavily streaked and difficult to separate from young Yellow Bitterns. They frequently get disoriented and fly into buildings or windows.

Geese and Ducks

Geese and ducks are usually found on water except when indulging in strong direct flight with their necks fully extended. They usually have broad, flattened bills, long necks, and stout bodies which enable them to dive, dabble, or dip for food on open water. Their legs are generally short with webbed feet.

LESSER TREEDUCK

Dendrocygna javanica 41 cm (16″) Plate 2d

The only resident wild duck in Singapore, this bird has been recorded breeding at a number of locations, including Senoko, Poyan, and Sungei Buloh, and formerly at Marina South—a central city location. It is usually seen in small flocks swimming and diving on freshwater ponds and its numbers could increase in protected reserves with more of this habitat. It is extremely wary because of constant disturbance by trappers.

GARGANEY
Anas querquedula 41 cm (16″) Plate 2e

Although the Garganey is the most common migrant duck observed in Singapore, it occurs only in small numbers on secluded freshwater ponds, lakes, or marsh areas where it dabbles on the water for food. It is distinguished in non-breeding plumage by a broad white eyebrow, and in flight by a dark leading edge of the forewing. It also has a glossy green speculum wing patch.

COTTON PYGMY GOOSE
Nettapus coromandelianus 33 cm (13″) Plate 2f

This bird is attractive and of small size, with an unmistakable white head and neck, blackish crown, and dark green upper-parts extending as a collar around the lower neck. The female is less conspicuous, with browner plumage. In flight both sexes have white in the wings which is limited to a trailing edge for the female. The bird is mostly seen in small numbers on freshwater ponds at Poyan, but it also occurred at Marina South as a non-breeding visitor before the site was filled in.

Birds of Prey

Characterized by their predatory habits, birds of prey, or raptors, are superbly equipped with excellent eyesight, swift and strong acrobatic flight, long sharp talons, and strong hooked bills for grasping prey. Many predate on live flying animals (mostly other birds), on ground-based mammals such as rats and shrews, or on water creatures such as fish, frogs, and snakes. Some survive on waste scraps whilst others feed on dead carrion, sometimes found on roads.

OSPREY

Pandion haliaetus 56 cm (22″) Plate 3a

The Osprey occurs throughout the year although its numbers increase during November to March. It has a prominent, wide, dark brown band through the eye and across the upper breast and black patches on the underside of the wings. It is an expert hunter of live fish taken in spectacular swoops over sea waters. Nesting has not yet been confirmed in Singapore for this species but there are some summer records each year.

BLACK BAZA

Aviceda leuphotes 33 cm (13″) Plate 3b

A migratory species in Singapore, the Black Baza sometimes occurs in large flocks with some birds stopping over to rest and feed. At close quarters this attractive black-and-white raptor reveals a black crest and barred breast. In flight it resembles a crow in colour, size, and flight pattern. Its normal habitat comprises forests, but it may also be seen in many other rural habitats on migration.

BLACK-SHOULDERED KITE

Elanus caeruleus 33 cm (13″) Plate 3c

A resident raptor, mostly seen in open country areas, the Black-shouldered Kite is easily recognized from a distance by its hovering flight pattern. At rest its smart, predominantly white breast and distinct black shoulder on pale grey wings can be clearly observed, along with its red eye. It occurs throughout the island, including more urbanized areas such as Bedok, Marina South, and Clementi, but generally in small numbers. It is a solitary hunter and will readily alight in a tree or even on a lamp-post after catching prey that include small snakes, frogs, large insects, and small mammals. It appears tamer than most other raptors, often allowing a close approach.

BRAHMINY KITE

Haliastur indus 46 cm (18″) Plate 3d

The most common bird of prey in Singapore, the Brahminy Kite is equipped with a white head, neck, and upper breast plumage finely streaked with black that is discernible only at close range. The remaining plumage of the mature bird is a rich chestnut brown, providing easy separation from the White-bellied Sea-eagle. Essentially a scavenger, it is mostly seen feeding over water, either along the coastal areas or over inland reservoirs. Its nesting and roosting areas are found inland as well as in coastal areas, and may comprise any convenient group of high trees, even within housing estates, with casuarinas especially favoured. Immature birds are heavily streaked brown and bear little resemblance to adult birds, thus providing possible confusion with the BLACK KITE *Milvus migrans*, an uncommon migrant. This species is also larger in size (66 cm) and has a distinctive forked tail and long, angled wings.

WHITE-BELLIED SEA-EAGLE

Haliaeetus leucogaster 71 cm (28″) Plate 3e

The White-bellied Sea-eagle is of larger size than the previous species and has a distinctive longer neck and grey upper-parts. Its soaring flight and 'V'-shaped gliding pattern are prominent identification features. This impressive resident bird of prey is most often seen patrolling the coastal and inland waters searching for live fish and water snakes. The bird's large, untidy stick nest, constructed in a high tree or on a pylon is generally easily visible but, fortunately, not accessible. It often confirms its presence with a duck quacking call issued on return to its inland nest sites, which are generally located in the forest but occasionally found in close proximity to buildings, such as at Kent Ridge and Punggol. It is frequently harassed by crows, especially during nesting periods. Up to nine pairs have been recorded breeding in one year in Singapore.

JAPANESE SPARROWHAWK
Accipiter gularis 28 cm (11″) Plate 3f

The Japanese Sparrowhawk is a very common migrant that preys on other smaller birds using a remarkably adept and fast flight action to cause panic. Small in size but aggressive in nature, this species is well fitted for survival with a sharp yellow bill and legs. Distinguished by a long tail and rounded wings, it may be seen in all habitats.

Rails and Crakes

Secretive and shy birds, rails and crakes inhabit reed or grass swampy areas and use their long-toed feet to walk on water plants. They are generally of small size and drab plumage, although there are some notable exceptions. Their short wings and tails are used for the limited flight that they reluctantly take before they disappear, with long legs dangling, into deeper undergrowth. They are usually most active at dawn and dusk, spending the rest of the day hidden in the undergrowth where they nest near the ground.

SLATY-BREASTED RAIL
Rallus striatus 25 cm (10″) Plate 4a

A distinctive chestnut crown, white throat, and white barred upper-parts serve to distinguish this species from other similar rails. Although a common resident, it is not frequently seen because of its secretive habits. It occurs wherever suitable water marsh habitat exists, including wet grass and streams with associated long vegetation on wasteland sites, and even in the centre of urban developments such as Orchard Road. The RED-LEGGED CRAKE *Rallina fasciata* is rarely seen but may be more common than supposed. It has distinctive red legs, reddish brown upper-parts, and white barring on wing primaries and lower breast, and occupies woodland habitats more than other closely related species.

WHITE-BROWED CRAKE

Porzana cinerea 20 cm (8″) Plate 4b

A dull brown bird with a grey head and breast, the White-browed Crake bears unmistakable white streaks on either side of a black eye stripe. It is found only in freshwater marsh areas such as Kranji Reservoir, Sungei Buloh, and Poyan, and is more easily viewed when flushed out of long vegetation. Although similar in habit to previous species, it is not so widely distributed.

WHITE-BREASTED WATERHEN

Amaurornis phoenicurus 33 cm (13″) Plate 4c

This is a comical, perky bird with a ghostly appearance when seen face on because of its green bill, red eyes, and white face and upper breast. The rear view comprises a rust-coloured rump often jerked in agitation when disturbed. It generally feeds on the ground in watery habitats but occasionally climbs into low shrubs to feed on seeds. A very common resident seen in all parts of the island, it often occurs along roadsides and is present in most parks and gardens.

WATERCOCK

Gallicrex cinerea 44 cm (17″) Plate 4d

The Watercock is a migrant visitor occurring from September to March; therefore, the full breeding plumage of the male, shown here, is almost never seen in Singapore. In non-breeding plumage this bird portrays its lighter brown colour and has green bill and legs. A larger bird than the previous species, it generally stays hidden in the undergrowth close to reservoirs and other freshwater marsh areas. In flight its neck and legs are outstretched. It generally flies further than the smaller rails when disturbed.

COMMON MOORHEN

Gallinula chloropus 33 cm (13″) Plate 4e

A common resident mostly seen on ponds and lakes, the Common Moorhen is easily identified by its attractive red shield to the head and black-and-white plumage. It swims along the water with a distinctive head-nodding action and feeds on water insects and plant life but never dives. It requires thick vegetation as a refuge for protection and breeding and is therefore absent from most of the larger reservoir areas without edge vegetation.

PURPLE SWAMPHEN

Porphyrio porphyrio 43 cm (17″) Plate 4f

A large bird, the size and shape of a chicken, the Purple Swamphen has an attractive purplish-blue plumage and red legs. The head and neck on some individuals are whitish but all have a large red bill. It is a common resident in the better marsh areas at Kranji, Poyan, and Tuas. As with other family members, it is reluctant to fly great distances but is less secretive and is often seen feeding on water plants on which it walks using its long, extended toes.

Shorebirds

These are generally known as waders. Most species are migratory and fly enormous distances each year between breeding areas in the north and non-breeding areas in the south. Good numbers of waders are attracted to Singapore's shores because of the country's location, as a principal land-based routing from north to south, at the tip of the Malay Peninsula. There are 45 species of wader on the current Checklist. The best periods for observation are from March to May and from September to November.

Waders are generally dull brown and white in non-breeding plumage when seen in Singapore and it is difficult to identify closely related species. A telescope is recommended for observation because of the problems of getting close. Closer inspection, however, reveals different body sizes and shapes. Bill and leg length and colour are also useful identification points. Feather colour and patterns are less definitive and are therefore more difficult to distinguish except by experts.

Shorebirds in Singapore may be found on expansive mud-flats offshore, mainly along the northern coast of the main island or, more easily, on mud-flats within tidal prawn ponds or inland river estuaries.

Some species prefer sandy shores and may be found along the eastern Changi coastline.

There is only one wader species resident in Singapore, the Greater Paintedsnipe, which is relatively uncommon.

GREY PLOVER

Pluvialis squatarola 28 cm (11″) Plate 5a

This is a comparatively large plover with greyish plumage. It stands taller, having longer legs than other waders. It is also less flighty and busy and is most often seen in small numbers, but usually singly, on sandy shorelines at Changi. In flight the characteristic black armpits under the wings may be observed.

PACIFIC GOLDEN PLOVER

Pluvialis fulva 25 cm (10″) Plate 5b

One of Singapore's most common migrant waders, the Pacific Golden Plover can be seen in large flocks at most mud-flat and sandy locations. The golden-and-black spangled upper-parts are more prominent closer to the breeding periods, in April and May. Some birds may also attain black breasts at this time. In the non-breeding plumage, the upper-parts are pale brown, spotted, and edged with yellow. The bird stands tall and slim, and may remain stationary when threatened. It feeds with a stopping and running action, and roosts on mud bunds or rocks when the tide rises.

LITTLE RINGED PLOVER

Charadrius dubius 18 cm (7″) Plate 5c

The Little Ringed Plover is fairly common but is much smaller than previous related species. It is distinguished by yellow legs, a dark breast band, and black head markings with a yellow eye ring. Generally very active, it adopts a run–stop technique when feeding on mud-flats, but also frequents firmer sandy beach areas. A closely related species, the COMMON RINGED PLOVER *Charadrius hiaticula*, is less common. It is slightly larger, has orange legs and an orange bill, and a white wing bar that is absent in the Little Ringed Plover.

KENTISH PLOVER

Charadrius alexandrinus 15 cm (6″) Plate 5d

This species is a common passage migrant usually found in small numbers on the sandy shorelines at Changi. The male lacks the sandy head acquired during breeding but is distinguished by an incomplete brown breast band, light under-parts, dark grey legs, and a black bill. It is slightly smaller than previous species and adopts a hunched stance.

MALAYSIAN PLOVER

Charadrius peronii 15 cm (6″) Plate 5e

This bird is similar to the Kentish Plover but the male has a black hind neck collar that extends to the sides of the breast, and sandier upper-parts. It also has a smaller bill and longer legs. The female has a more sandy rufous plumage and lacks the black collar. It is an attractive bird which occurs in small numbers along the Changi coastline during the winter. It breeds close by in Malaysia.

MONGOLIAN PLOVER

Charadrius mongolus 20 cm (8″) Plate 5f

The Mongolian Plover is slightly larger than previous species and occurs in large numbers, especially on mud-flats. It has a squat appearance, a distinctive white forehead, and black around the eye that develops as a stripe in breeding plumage. It lacks the white collar on the hind neck commonly found in other plovers. The closely related GREATER SAND-PLOVER *Charadrius leschenaultii* is slightly larger and taller (23 cm) with a heavier bill and lighter greenish-yellow legs (not black), although mud splattering can cause confusion. Generally, it is much less common in Singapore than the Mongolian Plover.

EURASIAN CURLEW

Numenius arquata 58 cm (23″) Plate 6a

The Eurasian Curlew is a very large shorebird with a long, decurved bill. It is much less common than the Whimbrel in Singapore and often occurs singly. In flight it has a distinctive white 'V' on the rump. It is mainly seen on mud-flats but is extremely wary, and frequently issues a distinctive 'coorlee' call which serves to warn other waders of danger. A closely related species, the EASTERN CURLEW *Numenius madagascariensis*, is very similar but slightly darker and lacks the white rump in flight. This species is a rare visitor to Singapore.

WHIMBREL

Numenius phaeopus 43 cm (17″) Plate 6b

Although of similar appearance to the previous species, the Whimbrel is smaller with a shorter bill that is more curved to the tip. It also has a distinctive striped crown. A regular visitor, it occurs in flocks of up to 100 birds, and is usually seen on mud-flats and along river estuaries of the northern shores. It is particularly vocal, with a distinctive whinnying call and a harsh screech.

BLACK-TAILED GODWIT

Limosa limosa 41 cm (16″) Plate 6c

The Black-tailed Godwit is a smart-looking, tall, and slender wader with a long, straight black bill with pinkish base. It occurs in small numbers on passage and is most often found feeding with other smaller waders. In non-breeding plumage it lacks orange to the breast and usually has a greyish upper breast. The closely related species, the BAR-TAILED GODWIT *Limosa lapponica*, is much less common and smaller (38 cm) with a shorter bill. It is distinguished by a barred rather than a black end-tail pattern in flight and lacks the white wing bar on the upper-parts.

COMMON REDSHANK

Tringa totanus 28 cm (11″) Plate 6d

A very common migrant, the Common Redshank is readily distinguished by its bright red legs and straight black-and-orange, but sometimes all black, bill. In flight it displays a white trailing edge and 'V'-shaped rump. Large numbers flock together and busily feed on mud-flats. A distinctive 'teu-hu-hu' call is issued when disturbed. Another wader of slightly smaller size, with red legs, is the TEREK SANDPIPER *Xenus cinereus*. This species is less common, has a more

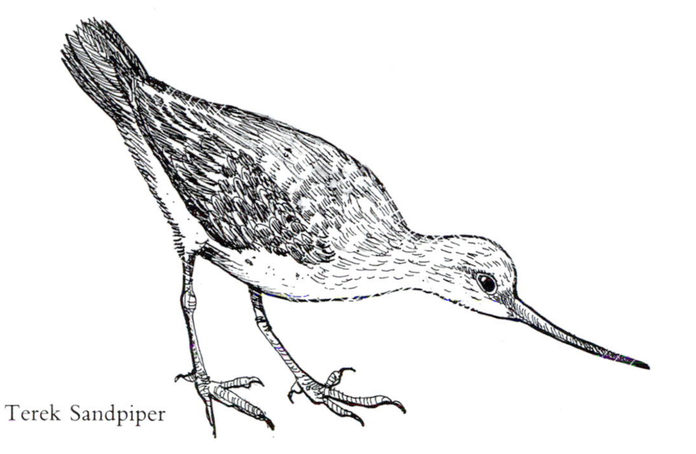

Terek Sandpiper

squat appearance, and possesses a distinctive long, slightly upturned bill. A very active and erratic feeder, it can usually be seen dashing through shallow water in both sandy and mud-flat locations.

MARSH SANDPIPER
Tringa stagnatilis 25 cm (10″) Plate 6e

A very common wader of delicate proportions, the Marsh Sandpiper has a small white head, long neck, and long green legs. It may be found in freshwater habitats as well as tidal mud-flats, where it adopts a distinctive sweeping bill action when feeding. The COMMON GREENSHANK *Tringa nebularia* is similar but larger (36 cm) and stouter, with a thicker, slightly upturned bill. It often occurs singly but feeds with other waders in similar habitats. It may also be seen in freshwater reservoir areas, but is much less numerous than the Marsh Sandpiper.

COMMON SANDPIPER
Actitis hypoleucos 20 cm (8″) Plate 6f

The Common Sandpiper can be observed throughout the year although it is a migrant and winter visitor. Identified in flight by a white wing bar and stiff wing beats, it has a pale eyebrow stripe and eye ring, and often bobs its tail whilst feeding. It usually occurs singly in both freshwater and tidal coastline habitats and along inland concrete water culverts or around park ponds. The WOOD SANDPIPER *Tringa glareola* is less common but often occurs in small flocks and has a more upright stance. The distinctive white speckling to its upper-parts, and the lighter plumage, white upper tail coverts, lack of wing bar, and conspicuous white eye stripes are good field marks.

PINTAIL SNIPE

Gallinago stenura 25 cm (10″) Plate 7a

One of a number of similar species, the Pintail Snipe is distinguished by an extremely long, heavy, straight bill and a striped brown-and-white head. It prefers swampy grass areas but may also occur on mud-flats where good cover is available to camouflage its subtle brown, white-streaked plumage. It was formerly a popular gamebird because of its twisting, rapid flight. During migration it occurs in large numbers throughout the island wherever suitable habitats exist. The COMMON SNIPE *Gallinago gallinago*, which is also a common migrant, is very similar but has a white trailing edge to secondaries of the wings, although this is not always easy to see. It usually has a more erratic zigzag flight when flushed out.

RUFOUS-NECKED STINT

Calidris ruficollis 17 cm (6½″) Plate 7b

A very small wader, and the most common of three similar species that occur in Singapore, the Rufous-necked Stint has a black bill and legs and paler grey upper-parts compared to other stints. In the non-breeding plumage, usually seen locally, there is no rufous on the neck. The bird occurs in large numbers, mixing with other mud-flat waders, and has a busy probing action in feeding. The LONG-TOED STINT *Calidris subminuta* is less common but occurs in the same flocks. It has yellow legs with heavy brown-and-black scaled upper-parts. The TEMMINCK'S STINT *Calidris temminckii* is the least common of the three, and can be distinguished by its unstreaked grey breast, darker brownish-grey upper-parts of uniform colour, and yellowish or greenish legs.

CURLEW SANDPIPER

Calidris ferruginea 22 cm (8½″) Plate 7c

The most common of the migrant sandpipers in Singapore, the Curlew Sandpiper is distinguished by its white eyebrow stripe, relatively long, slightly decurved black bill, and black legs. In flight it has a white rump and narrow white wing bars. The rufous breast of the breeding plumage is regularly seen at the start of the migration season in Singapore. It prefers tidal mud-flats where large flocks constantly feed with a pecking and probing action. The BROAD-BILLED SANDPIPER *Limicola falcinellus* is much less common and noticeably smaller, with a broader but shorter bill slightly decurved to the tip. Its most obvious feature is a double white eyebrow stripe and a white 'V' on the upper-parts similar to snipe plumage.

Terns

Predominately white or grey sea birds, terns have long, narrow wings and pointed bills. They are usually seen in flocks, hovering, diving, or gliding over the waters but are rarely seen swimming.

WHITE-WINGED TERN

Chlidonias leucopterus 26 cm (10″) Plate 7d

A very common migrant, the White-winged Tern prefers to feed by a gentle gliding and dipping action across the surface of inland freshwater lakes and coastal waters. It is seen mostly in its winter plumage, but in April or May it may acquire a black breast and reddish bill and legs. It is difficult to distinguish from the WHISKERED TERN *Chlidonias hybrida*, which is slightly larger (28 cm) and much rarer, although it may occur in larger numbers than previously thought. This species has a prominent white forecrown and 'V'-shaped black band to the nape and a grey rather than white plumage, especially to the rump.

LITTLE TERN

Sterna ablifrons 23 cm (9″) Plate 7e

The Little Tern is the smallest tern and has a brilliant white plumage. Although commonly seen along the east coast at Changi and northern shores at Seletar, where it has established breeding status in recent years, it is also often seen diving for fish at Kranji Reservoir and along coastal waters. The BLACK-NAPED TERN *Sterna sumatrana* also breeds in Singapore on a rocky islet near Changi, but rarely comes inland. It has a prominent black band around the nape and a deeply forked tail.

GREAT CRESTED TERN

Sterna bergii 46 cm (18″) Plate 7f

The Great Crested Tern is of large size, with a heavy greenish-yellow bill, a black crest which is sometimes raised, and a white forehead. It may be seen all year round but sightings decrease during the breeding season. It usually patrols deeper waters and frequently uses offshore *kelong* poles for perching, most often in the Straits of Johore. It is rarely seen inland. The LESSER CRESTED TERN *Sterna bengalensis* is slightly smaller (38 cm), more slender, and has an orange bill and a black forehead in the breeding season.

Pigeons and Doves

Despite the two names, there is no scientific difference between pigeons and doves as they form part of the same family group. There is a tendency, for aesthetic reasons, to refer to those species which have pastel shades of grey, brown, or pink and have 'soft' feathers as doves, but there are exceptions.

Pigeons and doves are a very well-known and adaptable family of birds. They are able to survive in most urban habitats, except for the specialist feeders who generally depend on seeds or fruits. They have very short legs which give the appearance of sitting on the ground, small heads and plump bodies, and a strong direct flight. For some species their narrow stiff wings are held in 'V' formation in flight, resembling a paper aeroplane gliding technique. Many species will fly long distances to feed when figs and berries ripen.

THICK-BILLED PIGEON

Treron curvirostra 27 cm (10½″) Plate 8a

The male is a remarkably beautiful bird with striking maroon back and green breast colouring. Both sexes have a pronounced green eye ring and a yellow bill with a red base. The Thick-billed Pigeon is not common in Singapore, and is only seen in larger numbers when fruit trees in season attract visitors from Malaysia.

PINK-NECKED PIGEON

Treron vernans 27 cm (10½″) Plate 8b

This bird is extremely common and occurs in large flocks, especially in the Central Catchment area, but it may also be seen in most parks and gardens away from the forest. The female has a dull green breast and green head. It prefers higher trees like casuarinas for communal roosting and preening, but also feeds in low shrubs such as *Simpoh Ayer*. Occasionally, it is seen on the ground drinking from water pools. Nests comprise a large collection of untidy sticks. The bird's call resembles a bubbling 'coo-coo'.

RED TURTLE-DOVE

Streptopelia tranquebarica 23 cm (9″) Plate 8c

This smart, handsome species, with its well-defined features, has increased in numbers since its introduction in recent years. It is still more restricted to rural habitats than the following species and the best places to see good numbers perched on fences or feeding on short grass are the Changi Airport perimeter road and the Punggol grasslands area. It is now known to be breeding in Singapore.

SPOTTED DOVE

Streptopelia chinensis 30 cm (12″) Plate 8d

The Spotted Dove is a common resident throughout the island, and can readily be identified by its large black neck patch with white spots.

Peaceful Dove

Its long tail has white outer feathers most easily seen in flight. It feeds on the ground, including hard pavings and road edges, and makes a loud wing-clapping sound when taking flight. It frequently issues a soft 'cuck-coo-coo' call and is particularly sociable, breeding throughout the year. The PEACEFUL DOVE *Geopelia striata* is noticeably smaller, more slender, and more attractive with black-and-white barring to the sides of its delicate neck and breast. Although it is basically a rural bird, it still occurs in good numbers. It is locally known as a 'merbok'. Its distinctive soft trill call, with a series of short 'coos', has made it one of the most popular cage-birds in Singapore.

GREEN-WINGED PIGEON

Chalcophaps indica 25 cm (10″) Plate 8e

Also called the Emerald Dove, this attractive species spends most of its time in the lower shrub of the forest. It is most often seen when flushed out of this habitat or flying across trails, but occasionally comes out to feed on forest tracks in the early morning. Its plumage colouring is unmistakable. It issues a low, mournful 'cu-oo' call and is often found in the Central Catchment area.

Parrots

Parrots are amusing, attractive birds with strong hooked bills and colourful plumage. They are swift, direct fliers with narrow, pointed wings. Gregarious and adept climbers, they use their feet to hold food. Parrots are very popular cage-birds, and many species in Singapore are escapees.

LONG-TAILED PARAKEET

Psittacula longicauda 41 cm (16″) Plate 8f

This is a beautiful bird with long, blue, pointed tail feathers seen clearly as it flies swiftly overhead in small flocks issuing a distinctive screech call. It often uses its strong claws to feed on fruit buds and to clamber about on branches. Although seen throughout the island it prefers forested areas where it nests in tree holes.

Cuckoos

Cuckoos are renowned for their parasitic habit of laying eggs in other birds' nests although not all species in this family have adopted this habit. Many species migrate through Singapore and are exciting but difficult birds to find because of their elusive nature. Learning their calls is particularly helpful in identifying their presence and in distinguishing various species. Most cuckoos are generally of greyish or greenish colouring and many have attractive breast barring. The most common feature is a long, graduated tail and a hawk-like flight. Once located, these migrants may be closely approached.

INDIAN CUCKOO

Cuculus micropterus 33 cm (13″) Plate 9a

The Indian Cuckoo is a relatively common migrant of similar appearance to several other cuckoos in the cuculus family. Most individuals have a yellow eye ring. The narrow black-and-white barring to the lower breast is also a distinctive feature. It issues a distinctive four-note call, 'ko-ko-ta-ko', of constant pitch except for the lower last note. Although it is most often found in the catchment areas, it also appears regularly on Sentosa, St. John's, Kent Ridge, and other secondary woodland locations.

BANDED BAY CUCKOO

Cacomantis sonneratii 23 cm (9″) Plate 9b

This is a resident species but is not common nor often seen. Again, the call is distinctive, varying from a four-note whistle, 'smoke yor-pepper', to a slower rising call, 'toe-tee-toh-toy', repeated several times and concluding with a chatter. Small in size, the Banded Bay Cuckoo is most likely to be found in secondary forest, although it may also be seen in rural open country areas and large gardens. It parasitizes the nests of the tailorbirds and Common Iora, and generally keeps to the lower shrubs.

PLAINTIVE CUCKOO

Cacomantis merulinus 22 cm (8½″) Plate 9c

Another rare resident, the Plaintive Cuckoo has none of the thick barring commonly found in other cuckoos but has a chestnut lower breast with a grey throat and head. More often encountered in open country areas on the offshore islands of Ubin and Tekong, it may be seen on the treetops from where it will call a plaintive, mournful, accelerating 'tay-ta-tee' descending rapidly in conclusion. It often continues to call throughout the night and in heavy rain. It parasitizes similar species to the Banded Bay Cuckoo.

COMMON KOEL

Eudynamys scolopacea 43 cm (17″) Plate 9d

An increasingly common migrant, the male is totally black with a prominent red eye and green bill, but is rather elusive and difficult to see. The female has attractive white speckles and streaks on a dark brown plumage. The bird is easy to identify by its loud 'ko-el' or 'ka-wao' call, which rapidly accelerates and intensifies. (The Black-naped Oriole has a similar call but this does not accelerate.) Its presence is recorded from many habitats, including Sungei Buloh, Kent Ridge, and the offshore islands of St. John's and Sentosa. It seems to be less common in forest and is more likely to be found in open country.

CHESTNUT-BELLIED MALKOHA

Phaenicophaeus sumatranus 41 cm (16″) Plate 9e

This is an unmistakably attractive bird with a large, apple green bill, a red eye patch, emerald green wings, and grey under-parts—except for the chestnut belly that is not always obvious. The only malkoha left in Singapore and an uncommon resident, it is highly dependent on forest habitation. Although regularly seen in the Central Catchment area and along pipeline trails, it is rarely located elsewhere.

GREATER COUCAL

Centropus sinensis 53 cm (21″) Plate 9f

An enormous black-and-tan bird that frequents larger gardens as well as forested areas, the Greater Coucal resembles a cross between a crow and a pheasant. It possesses a large, strong bill used to catch snakes and lizards as well as large insects. It issues deep, hollow, hoop notes repeated many times, and usually keeps close to the ground or clambers about awkwardly in low trees or shrubs. The smaller (38 cm) but similar LESSER COUCAL *Centropus bengalensis* is usually distinguished by white streaking to the plumage and a chestnut wing lining, but this is not easy to identify. More commonly found in open grasslands or wasteland areas, its call comprises a monotonous 'curra-currah-currah' resembling water pouring from a full bottle.

Owls and Nightjars

All species in this family are most active at dusk, dawn, or during the night. Their predominant features comprise large eyes for dark conditions, silent flight to surprise prey, and mottled plumage for excellent camouflage when roosting or nesting during the day. The owls are considered to be unlucky in local culture and many people are frightened by their moon-shaped facial discs, sharp bill and talons, and ghostly calls. The Nightjar's enormously wide gape allows it to consume large quantities of insects in flight, mainly during dusk forays. It is therefore an important participant in natural pest control.

BARN OWL

Tyto alba 34 cm (13½″) Plate 10a

A very distinctive owl with a white, heart-shaped facial disc, white breast, and grey and light brown upper-parts, the Barn Owl is thought to be fairly common in Singapore, although it is only regularly seen at known nesting spots such as Benjamin Sheares Bridge and Wessex Estate, and at night in the Changi and Loyang areas where it hunts. Its numbers have probably increased in recent years with overspill populations from the Malaysian palm oil plantations, where the owl was introduced to control the rat population. It issues a loud screech when disturbed. The COLLARED SCOPS-OWL *Otus bakkamoena* is much smaller (23 cm) and has yellow eyes and small ear tufts. It is lightly mottled brown with a paler collar on the back of the neck. It is a common resident in urban areas and forests, and its presence is usually confirmed by a characteristic mellow-toned double hoot, 'wo-ok', but it is not often seen.

BROWN HAWK-OWL

Ninox scutulata 30 cm (12″) Plate 10b

The Brown Hawk-owl is more hawk-like than other owls, with a small head, large yellow eyes, and a white breast heavily streaked with brown. It is also a resident species but is more restricted to forests than the previous species. It confirms its presence with an amusing 'hoo wup' call, which accelerates on the last syllable. It feeds on large insects, including cicadas, roaches, and dragon-flies, and also small mammals. The SPOTTED WOOD-OWL *Strix seloputo* is Singapore's largest (48 cm) resident owl but is considered to be endangered, with perhaps only a few pairs left. It is most often heard issuing a low, dog-barking call, 'hoo hoo hoo',

Spotted Wood-owl

in the Botanic Gardens, the Turf Club, and the Central Catchment area.

LARGE-TAILED NIGHTJAR
Caprimulgus macrurus 30 cm (12″) Plate 10c

This species has a distinctive white throat and uses its mottled brown plumage to sit in camouflaged comfort on the ground during the day. It is the most common nightjar in Singapore occurring in many urban areas wherever trees are present and insects are in good supply. The call is a monotonous 'tock tock' issued at dusk and onwards throughout the night, much to the consternation of light sleepers. It may often be seen sitting on quiet dry roads or circling around lights with bats at dusk. It nests on the ground or on flat roofs where dry leaves or pine needles are present. The MALAYSIAN EARED NIGHTJAR *Eurostopodus temminckii* is rare and is mainly confined to the Central Catchment Nature Reserve, where it hawks for insects along the water courses and swampy areas, issuing an unmistakable three-note call, 'get more be-er'.

Swifts and Treeswifts

Swifts are long, slender-winged aerial feeders which attain high air speeds. They are usually gregarious. Their plumage is generally black or dark brown. Treeswifts are a separate family with more colourful plumage and deeply forked tails. They perch with their long wing tips crossed.

EDIBLE-NEST SWIFTLET
Aerodramus fuciphagus 13 cm (5″) Plate 10d

A common resident, the Edible-nest Swiftlet is seen in large numbers, especially at dusk, aerial feeding with stiff wing beats. It constructs the much sought after creamy white saliva nest cups used for making soup, but it is illegal to collect these in Singapore. The almost identical BLACK-NEST SWIFTLET *Aerodramus maximus* also occurs in Singapore, but in much smaller numbers, and cannot be distinguished in flight. The black feathers used in its nest construction are the only obvious factor to separate the species.

HOUSE SWIFT
Apus affinis 15 cm (6″) Plate 10e

The House Swift has a conspicuous white rump and throat patch which is seen in flight while it feeds on insects. It nests in colonies under pier constructions along the north coast and in older house eaves on the northern islands. Lees common than the previous species, most birds probably come from residential quarters in Malaysia. It may be confused with the FORK-TAILED SWIFT *Apus pacificus*, a passage migrant, which also has a prominent white rump but is larger and more streamlined, with a deeply forked tail. The ASIAN PALMSWIFT *Cypsiurus balasiensis* occurs in small numbers and breeds in the Botanic Gardens. The plumage is blackish-brown with no white. The tail has a small fork.

GREY-RUMPED TREESWIFT
Hemiprocne longipennis 20 cm (8″) Plate 10f

An unusually attractive species, the Grey-rumped Treeswift is best identified by its long, pointed tail with streamers held closed in flight. It constructs an incredibly small saliva nest cup, precariously balanced on a horizontal tree branch, in which one egg is laid. Locally, it is common in wooded areas and forests, including Bukit Timah, Kent Ridge, the Turf Club, and in larger parks and gardens.

Kingfishers

Kingfishers are remarkably colourful birds with large, heavy bills, short tails and wings, and plump bodies. Capable of fast acrobatic flight, they are also noisy and active. They may feed on large insects as well as fish, and usually perch with their short legs on a suitable post or branch to watch and dive for prey. In Singapore, there are eight listed species, five of which are residents.

1. (a) Grey Heron, p. 21. (b) Purple Heron, p. 22. (c) Little Heron, p. 22. (d) Cattle Egret, p. 23. (e) Great Egret, p. 23. (f) Pacific Reef-egret, p. 23.

2. (a) Black-crowned Night-heron, p. 24. (b) Yellow Bittern, p. 24. (c) Cinnamon Bittern, p. 25. (d) Lesser Treeduck, p. 25. (e) Garganey, p. 26. (f) Cotton Pygmy Goose, p. 26.

3. (a) Osprey, p. 27. (b) Black Baza, p. 27. (c) Black-shouldered Kite, p. 27. (d) Brahminy Kite, p. 28. (e) White-bellied Sea-eagle, p. 28. (f) Japanese Sparrowhawk, p. 29.

4. (a) Slaty-breasted Rail, p. 29. (b) White-browed Crake, p. 30. (c) White-breasted Waterhen, p. 30. (d) Watercock (breeding), p. 30. (e) Common Moorhen, p. 30. (f) Purple Swamphen, p. 31.

5. (a) Grey Plover, p. 32. (b) Pacific Golden Plover, p. 32. (c) Little Ringed Plover, p. 32. (d) Kentish Plover, p. 33. (e) Malaysian Plover, p. 33. (f) Mongolian Plover, p. 33.

6. (a) Eurasian Curlew, p. 33. (b) Whimbrel, p. 34. (c) Black-tailed Godwit, p. 34. (d) Common Redshank (breeding), p. 34. (e) Marsh Sandpiper, p. 35. (f) Common Sandpiper, p. 35.

7. (a) Pintail Snipe, p. 36. (b) Rufous-necked Stint (breeding), p. 36. (c) Curlew Sandpiper, p. 37. (d) White-winged Tern, p. 37. (e) Little Tern, p. 38. (f) Great Crested Tern, p. 38.

8. (a) Thick-billed Pigeon, p. 39. (b) Pink-necked Pigeon, p. 39. (c) Red Turtle-dove, p. 39. (d) Spotted Dove, p. 39. (e) Green-winged Pigeon, p. 40. (f) Long-tailed Parakeet, p. 41.

9. (a) Indian Cuckoo, p. 41. (b) Banded Bay Cuckoo, p. 42.
(c) Plaintive Cuckoo, p. 42. (d) Common Koel, p. 42.
(e) Chestnut-bellied Malkoha, p. 43. (f) Greater Coucal, p. 43.

10. (a) Barn Owl, p. 44. (b) Brown Hawk-owl, p. 44. (c) Large-tailed Nightjar, p. 45. (d) Edible-nest Swiftlet, p. 45. (e) House Swift, p. 46. (f) Grey-rumped Treeswift, p. 46.

11. (a) Common Kingfisher, p. 47. (b) Stork-billed Kingfisher, p. 47. (c) Ruddy Kingfisher, p. 47. (d) White-throated Kingfisher, p. 48. (e) Black-capped Kingfisher, p. 48. (f) Collared Kingfisher, p. 48.

12. (a) Blue-tailed Bee-eater, p. 49. (b) Blue-throated Bee-eater, p. 49. (c) Dollarbird, p. 49. (d) Coppersmith Barbet, p. 50. (e) Mangrove Pitta, p. 53. (f) Pacific Swallow, p. 53.

13. (a) Rufous Woodpecker, p. 51. (b) Laced Woodpecker, p. 51. (c) Banded Woodpecker, p. 51. (d) Common Goldenback, p. 52. (e) White-bellied Woodpecker, p. 52. (f) Brown-capped Woodpecker, p. 52.

14. (a) Pied Triller, p. 54. (b) Ashy Minivet, p. 54. (c) Common Iora, p. 55. (d) Lesser Green Leafbird, p. 55. (e) Yellow Wagtail, p. 64. (f) Richard's Pipit, p. 65.

15. (a) Straw-headed Bulbul, p. 56. (b) Yellow-vented Bulbul, p. 56. (c) Olive-winged Bulbul, p. 56. (d) Greater Racket-tailed Drongo, p. 57. (e) Black-naped Oriole, p. 58. (f) Asian Fairy-bluebird, p. 58.

16. (a) Abbott's Babbler, p. 59. (b) Striped Tit-babbler, p. 59. (c) Magpie Robin, p. 60. (d) Asian Brown Flycatcher, p. 63. (e) Pied Fantail, p. 63. (f) Asian Paradise Flycatcher, p. 64.

17. (a) Flyeater, p. 60. (b) Oriental Reed-warbler, p. 61. (c) Common Tailorbird, p. 61. (d) Ashy Tailorbird, p. 62. (e) Yellow-bellied Prinia, p. 62. (f) Brown Shrike, p. 66.

18. (a) Philippine Glossy Starling, p. 66. (b) Black-winged Starling, p. 67. (c) Common Myna, p. 67. (d) White-vented Myna, p. 67. (e) Hill Myna, p. 68. (f) Large-billed Crow, p. 58.

19. (a) Brown-throated Sunbird, p. 68. (b) Purple-throated Sunbird, p. 68. (c) Copper-throated Sunbird, p. 69. (d) Olive-backed Sunbird, p. 69. (e) Crimson Sunbird, p. 69. (f) Little Spiderhunter, p. 70.

20. (a) Orange-bellied Flowerpecker, p. 70. (b) Scarlet-backed Flowerpecker, p. 70. (c) Eurasian Tree Sparrow, p. 71. (d) Baya Weaver, p. 71. (e) Scaly-breasted Munia, p. 71. (f) Chestnut Munia, p. 72.

COMMON KINGFISHER

Alcedo atthis 18 cm (7″) Plate 11a

A common migrant and winter visitor, always found in close proximity to water, the Common Kingfisher is relatively small with a distinctive white neck patch and chestnut ear patch. Often seen perched on poles or projecting bars over water, it is a fast-flying expert catcher of small fish. It can be seen throughout the island at parks and gardens with pools or ponds, and along the coastline. The very similar BLUE-EARED KINGFISHER *Alcedo meninting*, a rare resident confined to forests, is usually only seen flashing quickly along freshwater streams in the Central Catchment. It is best identified by call, which is a higher pitched squeak than that of the Common Kingfisher. It also has a blue ear patch.

STORK-BILLED KINGFISHER

Halcyon capensis 37 cm (14½″) Plate 11b

A large attractive species with a huge red bill and brown head and ginger breast, the Stork-billed Kingfisher is always found close to water, usually along sea coasts and in the northern islands of Ubin and Tekong. It occurs less regularly at Kranji Dam mangrove and in the Central Catchment area. It frequently issues a loud, wailing cry.

RUDDY KINGFISHER

Halcyon coromanda 28 cm (11″) Plate 11c

A distinctive red bill and bright vinous rufous plumage provide spectacular colour in the mangroves to which this species is confined. A rare resident in Singapore, the bird is usually seen only on Tekong, but another sub-species occurs as an uncommon migrant. It is a very shy bird, and is thought only to feed on marine life. The ORIENTAL DWARF KINGFISHER *Ceyx erithacus* is tiny in comparison, but is beautifully coloured with a black mantle, red back, dark blue wings, yellow upper-parts, and red bill. It is also a rare migrant and is sometimes recorded as a result of flying into windows and stunning itself.

WHITE-THROATED KINGFISHER
Halcyon smyrnensis 28 cm (11″) Plate 11d

The White-throated Kingfisher is a common, prominent resident most often found in open country and coastal parks. It issues a distinctive whinnying call, and feeds principally on large insects, diving for them in grass or low shrubs. In Singapore, the species also has a white upper breast. It nests in holes excavated in earth banks.

BLACK-CAPPED KINGFISHER
Halcyon pileata 30 cm (12″) Plate 11e

A common winter visitor, this species has a beautiful plumage comprising a distinctive black cap, orange breast, purple-blue wings, and a large red bill. It is mostly seen along forest streams and near inland reservoirs or prawn ponds, as well as in parks and gardens wherever water is present. It sometimes calls like the White-throated Kingfisher but is often silent, and is extremely wary.

COLLARED KINGFISHER
Halcyon chloris 24 cm (9½″) Plate 11f

Singapore's most common resident kingfisher, its colour can range from blue-green to darker blue with a black edging to the distinctive white collar. It makes its presence known with a very noisy call, much like a coarse laugh. It is fairly aggressive towards other birds, especially during nesting when it excavates holes in earth banks or uses tree holes. Mostly it frequents sea coasts, inland parks, and open rural habitats where it feeds on fish or insects.

Bee-eaters

These beautiful, graceful fliers specialize in aerial feeding on insects, from which they return to a high tree perch. They are characterized by delicate, slightly decurved bills and long tail streamers. They are usually green in colour and have long, pointed wings.

BLUE-TAILED BEE-EATER

Merops superciliosus 30 cm (12″) Plate 12a

More characteristically described as brown-throated, this bee-eater is a September to April visitor. The species is easily seen due to its habit of perching in exposed locations, including overhead telephone wires and dead or leafless trees. Large flocks are sometimes recorded and may be seen in any open country location and around the reservoir areas.

BLUE-THROATED BEE-EATER

Merops viridis 28 cm (11″) Plate 12b

This is a common resident that arrives in March, as the above species leaves, and returns south after breeding. More attractive than the previous species, it has a rich chocolate head and nape contrasting with a deep blue throat. The remaining plumage is green except for the long blue tail with central streamers. The bird nests in holes in the ground or in banks of sand or loose soil. Breeding has been confirmed in the Changi area and Sentosa and in quarries at Bukit Timah and Ubin. It prefers to reuse undisturbed sites for breeding, but may be seen in any open country. It issues a pleasant 'berek berek' call, giving rise to its Malay name.

Rollers

Rollers are stout, aggressive birds with long, broad wings used in acrobatic flight and for aerial feeding. They have a noticeably large head and short tail.

DOLLARBIRD

Eurystomus orientalis 30 cm (12″) Plate 12c

A familiar and conspicuous resident and winter visitor, the Dollarbird has long wings with round silver patches, giving rise to its name. It is often seen perched on top of a high tree for long periods. Although its numbers increase substantially between October and April, resident birds are small in number. A very aggressive and noisy bird, it issues harsh rasping notes, especially when chasing off other birds. It nests in tree holes and feeds on flying insects which it catches expertly in acrobatic flight, despite its heavy body structure.

Barbets

Barbets are colourful forest birds with large, bristle-sided bills and short legs and tails. They are avid fruit eaters, and issue monotonous repetitive but distinctive calls.

COPPERSMITH BARBET
Megalaima haemacephala 15 cm (6″) Plate 12d

This species is identified by its stubby shape, colourful red-and-yellow head markings, and distinctive green-striped yellow breast. Red legs and a green back are also dominant features. It has a remarkably penetrating, monotonous 'poop' call, which is issued rapidly. It is a very active bird found in open country habitats, although not in great numbers. It is more commonly observed on offshore islands such as Sentosa and St. John's where it nests in tree holes. The related but larger (25 cm) RED-CROWNED BARBET *Megalaima rafflesii* has a highly coloured plumage comprising a blue throat, yellow cheek, and red crown. It is mainly confined to the Central Catchment and Bukit Timah areas, where it is attracted to fruiting figs when in season and where it nests in tree holes. Unfortunately, this species has become less common in recent years and is more often heard than seen. Its call comprises a series of 'poop' notes issued at three per second, all on the same note.

Woodpeckers

Woodpeckers are colourful birds with extraordinary abilities which enable them to cling horizontally to tree trunks and to excavate tree holes with their strong, chisel-pointed bills. Vulnerable to extinction with reductions to forests, many species have been lost in Singapore over the last fifty years. Those that remain bring colour, movement, and excitement to forest bird-watching. Woodpeckers possess well-designed strong feet, stiff tails, and a long sticky tongue to collect insects, all of which make them a joy to see in action.

RUFOUS WOODPECKER

Celeus brachyurus 25 cm (10″) Plate 13a

The Rufous Woodpecker's plumage is mainly dark rufous brown with black barring to the upper-parts which is only visible at close quarters. The male has a red cheek patch, but both adults have black bills. It is an uncommon resident that occurs mainly in the Central Catchment area but is occasionally seen in mangroves at Kranji and in some gardens close to forests. It feeds principally on insects picked off tree-trunks. It issues a raucous, descending, four-note 'kwee' call, often in flight.

LACED WOODPECKER

Picus vittatus 30 cm (12″) Plate 13b

A larger woodpecker with a light green plumage, red crown (male), blue cheek patch, brownish-yellow throat, and a characteristic scaly lower breast, the Laced Woodpecker occurs in more open habitats, especially in coconut plantations. It is fairly common in Ubin and is regularly seen in Changi and Sentosa; it is less frequently encountered at Kent Ridge. The call is a chattering laugh preceded by a sharp 'kyp' note.

BANDED WOODPECKER

Picus miniaceus 25 cm (10″) Plate 13c

The banding across the lower breast of this species is a distinctive feature, but is not always easy to see. The wings and head are rufous brown with a conspicuous yellow crest. It is the most common forest woodpecker in Singapore, but is also seen in larger parks, especially the Botanic Gardens. A noisy species, it issues a strident 'kwee' call, frequently repeated, and is always very active, working its way up and along tree-trunks searching for insects in the bark.

COMMON GOLDENBACK
Dinopium javanense 30 cm (12″) Plate 13d

A beautifully coloured bird with a golden yellow back, bright red rump, and striped black-and-white head. The male has a red-crested crown while both adults have a chequered black-and-white breast. Not as common in Singapore as its name implies, this bird is nevertheless seen regularly in preferred habitats of open woodland, large gardens, coconut plantations, and mangroves, especially in rural areas and kampong environments such as Pulau Ubin. It issues a loud, machine-gun, staccato call, descending in scale.

WHITE-BELLIED WOODPECKER
Dryocopus javensis 43 cm (17″) Plate 13e

The White-bellied Woodpecker is mainly black in colour with a black upper breast and white belly below that sometimes has a yellowish tinge. The most prominent feature is a bright red crest for both sexes, but the male also has a red forecrown and moustache. This is a large woodpecker only found in the forest. It is now very rare, and is possibly extinct in Singapore; it was last recorded in 1988 in the Central Catchment area.

BROWN-CAPPED WOODPECKER
Picoides moluccensis 13 cm (5¼″) Plate 13f

The plumage of this species is brown with white banding. A brown cap to the head distinguishes it from the very similar grey-capped species found in Malaysian forests. A common resident, the Brown-capped Woodpecker is usually seen in pairs busily feeding on tree-trunks and branches, issuing a 'trill' call. It is found throughout the island, especially in mangroves, parks, and gardens, but not usually in forests.

Pittas

A much revered family of birds because of their strikingly colourful appearance, rarity, and elusive nature, pittas are normally found on the ground in thick forest areas. They have distinctive calls, comprising short loud whistles, to which they will respond when mimicked. They are usually plump in size with long legs.

MANGROVE PITTA
Pitta megarhyncha 20 cm (8″) Plate 12e

The only resident pitta in Singapore, the Mangrove Pitta is endangered because of its limited habitat of mangroves. Its presence was recorded in recent years from Tekong but breeding has not been proven. It mainly feeds on crabs. The call is a disyllabic whistle 'ta-law' with a drawn-out second note. Another migrant species that usually arrives annually in small numbers is the HOODED PITTA *Pitta sordida*. It has a black head with a chocolate crown, bright green back and breast, and a red lower belly. The blue wings have large white patches which can be observed during flight. It is most often seen in the Bukit Timah Nature Reserve feeding on the forest floor. The BLUE-WINGED PITTA *Pitta moluccensis* is almost identical to the Mangrove Pitta, but has a shorter bill and a light buff band over the eye. It is another rare winter visitor, sometimes seen in gardens but more likely in forest or mangrove. It is particularly prone to flying into glass windows.

Swallows

Swallows are familiar and common birds in many parts of the world. Some migratory species travel long distances while others are sedentary. They are characterized by pointed wings and slender, graceful flight, usually closer to the ground than swifts and other aerial feeders.

PACIFIC SWALLOW
Hirundo tahitica 14 cm (5½″) Plate 12f

The Pacific Swallow has a grey-streaked breast and extensive red throat and forehead. It is a common resident seen throughout the island in large numbers perching on sticks or wires in pairs or in large flocks. It generally prefers feeding over water and nesting under bridges and dams. The similar BARN SWALLOW *Hirundo rustica* has a whiter breast and blue throat collar, distinguishing it from the Pacific Swallow.

It is a passage migrant, but despite this can be seen most months of the year speedily flying with its long tail streamers over hills and in urban areas. It occurs in large flocks at certain periods, using overhead wires and even buildings in housing estates to rest during long migration flights.

Cuckoo-shrikes and Minivets

Frequenting woodland habitats, these families provide a range of species exciting to find and to watch. The birds are always actively searching for insect prey, either on the foliage or in flight in the tree canopy. The cuckoo-shrikes generally have larger bills and are of grey, white, and black plumage, while minivets are mostly of brilliant colours, are slimmer, and have longer tails.

PIED TRILLER
Lalage nigra 18 cm (7″) Plate 14a

A distinctive black-and-white common resident, the Pied Triller is mostly seen in the upper tree canopy feeding on insects, especially caterpillars. It occurs in parks, gardens, and roadside trees, and has a distinctive undulating flight pattern similar to that of a woodpecker. The immature bird lacks the larger white wing patch and has a generally greyer plumage. It may be confused with the Purple-backed Starling, but it usually occurs singly or in pairs and not in flocks.

ASHY MINIVET
Pericrocotus divaricatus 20 cm (8″) Plate 14b

A migrant species that occurs annually in small numbers, the Ashy Minivet may be found in forested areas and large parks and gardens. It is similar in appearance to the above species but has plain dove grey upper-parts and nearly always occurs in small flocks. It generally keeps to upper tree canopies and favours casuarinas. The related SCARLET MINIVET *Pericrocotus flammeus* is a rare resident generally only seen in Bukit Timah forest where it is unmistakable because of its brillant red plumage. The females, in contrast, are bright yellow in colour.

Ioras and Leafbirds

These are canopy-feeding species which are difficult to locate because of their green and yellow plumage. All have distinctive musical calls and are very active insect feeders.

COMMON IORA

Aegithina tiphia (singapurensis) 15 cm (6″) Plate 14c

The male Common Iora in Singapore has a black cap, black back, two distinctive white wing bars, and a bright yellow breast, while the female has a rich green back. It is a common resident in parks and gardens and in its natural habitat of mangroves. It issues a wide collection of calls, including musical whistles and trills, and feeds principally on insects, especially caterpillars taken from the leaves in the mid-storey of trees. It normally nests in a tree fork where it constructs a saliva cup structure.

LESSER GREEN LEAFBIRD

Chloropsis cyanopogon 18 cm (7″) Plate 14d

This is an uncommon resident which is often difficult to see because of its beautiful leaf green plumage. It issues a distinctive four-note whistle likened to 'mer-ry christ-mas' with a higher third note. It has occurred more regularly in recent years, mainly in Bukit Timah and the Central Catchment areas. The GREATER GREEN LEAFBIRD *Chloropsis sonneratii* is a similar but slightly larger species (20 cm) and is now rarely seen in Singapore. The male lacks the yellow edging to the black throat of the previous species. The BLUE-WINGED LEAFBIRD *Chloropsis cochinchinensis* is also similar to the Lesser Green Leafbird; its bright green plumage makes it difficult to spot against the leaves. At close quarters, it has an unmistakable blue edging to the wings. More regularly seen and heard in recent years, it occurs in both secondary and primary forest areas.

Bulbuls

Bulbuls are a familiar family of birds in the region. They comprise many species but only a few are resident in Singapore. They are distinguished by their extremely vocal musical calls and active social behaviour. Of moderate size, they generally have fluffy yellow or brown-and-white plumage.

STRAW-HEADED BULBUL
Pycnonotus zeylanicus 29 cm (11½″) Plate 15a

This is a large bulbul with a remarkable song composed of loud bubbling calls often issued in duet. The straw yellow head is conspicuous along with the white streaking to the greenish back and the paler breast. A common resident in Pulau Ubin, it is very rarely seen elsewhere except at Senoko where a small resident population is present. Other sightings, probably of escapees, have been made in Changi and Upper Peirce Reservoir Park.

YELLOW-VENTED BULBUL
Pycnonotus goiavier 20 cm (8″) Plate 15b

A very common resident seen in most parks and gardens throughout Singapore, the Yellow-vented Bulbul is thought to originate from mangroves but is absent from thick forest habitat. The brown-and-white streaked head and blackish eye are more distinctive than the yellow vent. The bird is very vocal and highly conspicuous, especially when feeding on fruits, seeds, and berries, but it is also partial to insects which it may raid from spiders' webs. It often nests in low bushes close to buildings. The CREAM-VENTED BULBUL *Pycnonotus simplex* is slightly smaller but has distinctive white eyes. It is becoming very uncommon in Singapore, being found only in the Central Catchment and Bukit Timah forest areas.

OLIVE-WINGED BULBUL
Pycnonotus plumosus 20 cm (8″) Plate 15c

This common resident is more often found closer to the forest than the Yellow-vented Bulbul. It has an indistinctive plumage, with a lighter olive green edging to the wing, a red eye, and buff brown vent. The call is a more throaty song than that of the previous species, and is frequently issued. The RED-EYED BULBUL *Pycnonotus brunneus* is of similar plumage but slightly smaller, and it has a brighter red eye. It is much less common, occurring only in the forested areas of the mainland.

Drongos

Most species of drongo are glossy black with long, distinctive, forked tails and red eyes. Noisy and aggressive, they are acrobatic feeders, catching insects on the wing, and are often seen perching in exposed locations.

GREATER RACKET-TAILED DRONGO
Dicrurus paradiseus 33 cm (13″) Plate 15d

This is a magnificent and bizarre resident species whose long, twin-shafted tail with tiny end feathers is distinctive during the breeding season. It occurs in good numbers in the forested areas but is rarely seen elsewhere. It has a wide repertoire of calls and can imitate other birds, but it usually betrays itself with characteristic 'cheop' notes. The similar CROW-BILLED DRONGO *Dicrurus annectans*, a migrant occurring in coastal shrub and forests, has a much heavier and stouter bill and a shallow, forked tail. The BLACK DRONGO *Dicrurus macrocercus* regularly occurs on passage but in small numbers. It has a deeply forked tail, is entirely black, and feeds on insects found in long grasses. It may be observed sitting conspicuously on wires in open country areas such as Punggol and Poyan.

Orioles and Fairy-bluebirds

These birds are of medium size but robust shape. They occur in woodland habitats and feed on insects and fruit. Most have attractive and characteristic musical calls which are frequently issued.

BLACK-NAPED ORIOLE
Oriolus chinensis 27 cm (10½″) Plate 15e

A surprisingly common resident occurring throughout Singapore, the Black-naped Oriole is of distinctive bright yellow plumage, with a red bill and a black nape patch behind the eye. The immature bird is greener and lacks the nape patch of the mature bird. The species feeds on insects, fruit, and berries, and issues a pleasant musical repertoire of fluty notes. It flies with an unusual wing-flapping action, freezing in mid-flight. It frequents roadside trees in the centre of the city as well as gardens and forest areas.

ASIAN FAIRY-BLUEBIRD
Irena puella 25 cm (10″) Plate 15f

A remarkably attractive forest resident, the Asian Fairy-bluebird is generally confined to the Central Catchment and Bukit Timah areas. The female has an all blue-green plumage but also has a distinctive red eye. It is more often heard than seen, and has a repetitive 'be quick' call which gradually rises in tone. It often occurs in pairs in the tree canopy, actively feeding on fruit and insects.

Crows

These are large birds with heavy, powerful bills. Considered to be intelligent, they are generally noisy and gregarious. They will feed on a wide variety of prey, including carrion and the eggs and young of other birds.

LARGE-BILLED CROW
Corvus macrorhynchos 51 cm (20″) Plate 18f

A common resident, this bird is aggressive and constantly harasses other species, particularly birds of prey, during the nesting season. It occurs in all habitats and feeds on all kinds of waste. It is considered to be highly resourceful and cunning. The similar and equally pesky HOUSE CROW *Corvus splendens* is distinguished from the previous species by its slightly smaller size, shorter bill, and greyish neck plumage. It prefers coastal areas to forests, but is also considered a nuisance. It has the dubious distinction of being the only wild bird that is unprotected in Singapore and is thus regularly hunted.

Babblers

Babblers form an enormously large family of birds of which only six species naturally occur in Singapore. Most of these are forest birds and are characterized by short, rounded wings and tails. Generally, they are sedentary residents vulnerable to the loss of shrub cover and sensitive to excessive disturbance.

ABBOTT'S BABBLER

Trichastoma abbotti 17 cm (6½") Plate 16a

A secretive resident species, this bird has an indistinctive dun brown plumage, a whitish throat, and tawny flanks. More widespread than other family members, the Abbott's Babbler nevertheless occurs only in small numbers in parks and gardens and forested areas. It has a loud, penetrating, three- or four-note call, sounding like 'do-wait-for-me', and often likened to a human whistle. The similar SHORT-TAILED BABBLER *Trichastoma malaccense* is confined entirely to forests and is becoming an endangered species in Singapore. It is usually identified by a loud, clear, whistled song of six or more wavering notes which drop rapidly in pitch. It can be called in but is always difficult to see.

STRIPED TIT-BABBLER

Macronous gularis 13 cm (5¼") Plate 16b

Small but attractive, this species has a chestnut crown and a prominently black-streaked yellow breast. It is a common forest edge bird that also occurs in gardens where it inhabits the lower shrub, searching for insects, often in pairs or even a family group. It frequently calls with penetrating 'chonk' notes and low, grating, scolding notes. The CHESTNUT-WINGED BABBLER *Stachyris erythroptera* is an endangered forest species with an attractive blue eye ring, a grey head, and rich chestnut wings. It keeps to low undergrowth whilst issuing distinctive, lower-pitched, hollow 'poop' calls which descend in pitch.

Thrushes

These comprise a large family of birds which are mainly ground dwellers searching for insects in the undergrowth. Several species of the true thrush occur in Singapore, but only on migration. The robins who belong to this family are featured here. They are excellent songsters and therefore popular cage-birds.

MAGPIE ROBIN

Copsychus saularis 22 cm (8½″) Plate 16c

This is an easily recognized black-and-white bird formerly common in parks and gardens and kampong areas, where it is seen feeding on worms and insects on the ground. Because of its tameness, its popularity as a cage-bird, and to losses of suitable habitat, this species has diminished to exceedingly low numbers in Singapore. In recent years, it has been reintroduced to the Botanic Gardens, but it also occurs in other locations with extensive garden habitat, most commonly found on the offshore islands, including Ubin, Tekong, and St. John's. It is also regularly seen at Medway Park, Changi Hill, and the Kranji area. The WHITE-RUMPED SHAMA *Copsychus malabaricus* is a much larger bird (28 cm) than the Magpie Robin and is also a popular cage species. An attractive bird with an orange lower breast, black upperparts and tail, and a distinctive white rump, it issues a remarkably loud and musical song. Although very rare in Singapore, some birds are located in their preferred forest habitat at Mandai and in Pulau Tekong North.

Warblers

These are small, active insect feeders usually found in the tree canopy or in shrubs. They are often of plain, dull plumage and have thin bills.

FLYEATER

Gerygone sulphurea 9 cm (3½″) Plate 17a

Along with the flowerpecker, this is the smallest bird in Singapore. The plumage lacks distinctive markings but is generally olive green on the back and tail with a dull yellow breast. It is a common resident, frequently heard but not often seen, and is recorded from many areas of the island because of its distinctive, penetrating call, best described as a high-pitched, buzzy, musical wheeze with long descending phrases. It is especially common in mangroves. The ARCTIC WARBLER *Phylloscopus borealis* is a common migrant found in the higher tree canopy. It is distinguished by its long, pale white eyebrow stripe and a rather inconspicuous white wing bar against a dull greenish or brownish plumage.

ORIENTAL REED-WARBLER

Acrocephalus orientalis 19 cm (7½″) Plate 17b

Larger and more common than other reed-warblers in Singapore, the Oriental Reed-warbler has plain brown upper-parts, a long buff eyebrow, and whitish under-parts with tawny flanks. It is easily distinguished by its loud, harsh, grating calls issued from a perch in long grasses or reeds near water. PALLAS'S WARBLER *Locustella certhiola* is smaller (15 cm) and has broad black streaks to brown upper-parts and a white-tipped tail. The BLACK-BROWED REED-WARBLER *Acrocephalus bistrigiceps* (14 cm) has a prominent black crown stripe above a buffy supercilium and a plain brown back. The LANCEOLATED WARBLER *Locustella lanceolata*, which is similar to the Pallas but smaller (13 cm), has extensive streaking to the breast. All these reed-warblers are migrants and are frequently encountered at Kranji Reservoir, Punggol, and Poyan.

COMMON TAILORBIRD

Orthotomus sutorius 12 cm (4¾″) Plate 17c

A small, perky, common resident species, the Common Tailorbird is renowned for its intricate nest-building ability, which involves sewing two large leaves together to form a nest cup. It is more likely to frequent parks and gardens in urban areas than other similar species and may use potted plants on balconies for nesting. In the breeding season it has an elongated tail, which it holds vertically as it searches through low bushes for insects. Its call is a monotonous, repetitive 'chee-yap'. The similar, and perhaps more common resident, the

DARK-NECKED TAILORBIRD *Orthotomus atrogularis*, is more closely associated with forest edge habitats but also occurs in gardens. It can be distinguished from the Common Tailorbird by its yellow vent, darker neck collar markings, and more extensive chestnut crown. Its call is highly variable, ranging from 'kri-ri-ri' to 'zit-zit'; a 'prrp prrp' telephone call is also distinctive.

ASHY TAILORBIRD

Orthotomus sepium 11 cm (4½″) Plate 17d

A resident most commonly found in or near coastal mangroves, the Ashy Tailorbird has a greyish-green plumage and a chestnut crown, cheeks, and chin. Although difficult to see, it is easily heard issuing repetitive 'whee-chip' notes. The similar RUFOUS-TAILED TAILORBIRD *Orthotomus sericeus* is much less common. It often occurs in similar habitats but in more rural locations, especially on the offshore islands. It is distinguished by a whiter throat, cheeks, and under-parts and a distinctive long red tail.

YELLOW-BELLIED PRINIA

Prinia flaviventris 14 cm (5½″) Plate 17e

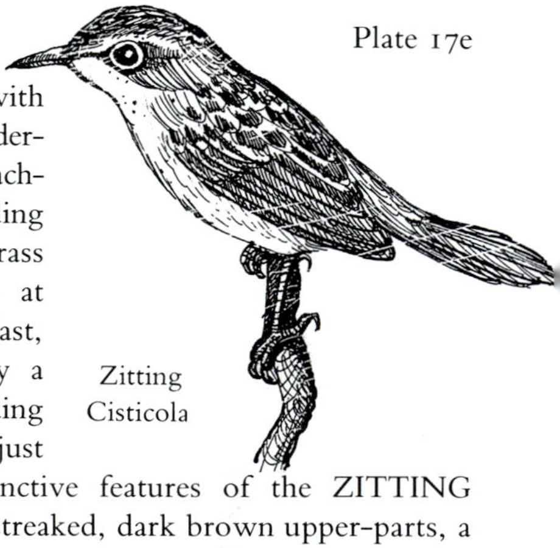

A small, slim, common resident, with a grey head, red eye, and yellow under-parts, this bird can be closely approached before it takes off with a bounding flight. It quickly occupies long grass in reclaimed wasteland habitats at Kranji, Punggol, and Marina East, where it is readily identified by a musical reeling call of descending notes which it issues from a perch just below the reed tops. The distinctive features of the ZITTING CISTICOLA *Cisticola juncidis* are streaked, dark brown upper-parts, a rufous rump, and a white-tipped tail. It is a small, short-tailed, stumpy, common resident occupying similar habitats in which it constructs a cup nest in long grasses. It is a very active bird with an erratic spiralling flight. It issues a continuous 'tik tik' call.

Zitting Cisticola

Flycatchers and Whistlers

There are a large number of species of this bird, many of which are migratory. Generally of small size with broad, flat bills designed for catching insects, they have a characteristic acrobatic flight from a favoured perch. The whistlers are heavier birds with large, round heads.

ASIAN BROWN FLYCATCHER
Muscicapa dauurica 13 cm (5¼″) Plate 16d

A common migrant of small, delicate proportions, the Asian Brown Flycatcher has an inconspicuous plain brown plumage with a prominent white eye ring. It has the usual upright stance of the species. It is most often encountered in the tree canopy, making forays for insects from a favourite branch. The closely related FERRUGINOUS FLYCATCHER *Muscicapa ferruginea* has rufous to the flanks and a lighter buff breast. It only occurs in very small numbers. At least one bird can be regularly seen in the Bukit Timah Nature Reserve from September to March.

PIED FANTAIL
Rhipidura javanica 18 cm (7″) Plate 16e

Mostly found in mangroves in Singapore where it is still a reasonably common resident, the Pied Fantail feeds close to the ground on insects. It is a very acrobatic and active bird with a beautiful, white-tipped fan tail exposed when perched. Friendly and inquisitive, it responds to imitations of its disorganized squeaky call. The MANGROVE BLUE FLYCATCHER *Cyornis rufigastra* is not closely related but occupies a similar habitat. It has beautiful plumage with blue upper-parts and a bright orange breast. A rare resident, suffering from the loss of mangrove forests to which it is confined, it is now found only on Pulau Tekong. The MANGROVE

Mangrove Whistler

WHISTLER *Pachycephala cinerea* is also an uncommon resident. Renowned for its indistinct, dull brown plumage, it has a more heavy chunky appearance than similar brown flycatchers. Also restricted to mangrove habitat, it occurs mainly on offshore islands, but may occasionally be found at Kranji, Pasir Ris, and Sungei Buloh. Its call comprises a loud whistle of three to four notes. The bird can be attracted in by imitation.

ASIAN PARADISE FLYCATCHER
Terpsiphone paradisi 22 cm (8¾″) Plate 16f

A beautiful migrant with a most distinctive white or rich rufous brown long tail and a crested dark blue head, blue bill, and eye ring, this bird occurs regularly in the secondary forest areas of the Central Catchment area on passage, but not in great numbers. A similar species, but one that is very rare, is the JAPANESE PARADISE FLYCATCHER *Terpsiphone atrocaudata*, which has much darker brown upper-parts and tail.

Wagtails and Pipits

These are terrestrial birds with long legs and tails, frequently seen feeding on insects on the ground. The wagtails have striking plumage but the pipits are mostly streaked with brown feathers. Most birds wag their tails and have an undulating flight which exposes their white outer tail feathers.

YELLOW WAGTAIL
Motacilla flava 18 cm (7″) Plate 14e

A common migrant, the Yellow Wagtail is seen on golf courses and other short-cut grass areas feeding on insects. Its non-breeding plumage is brown with yellow breast patches. When disturbed, it issues a characteristic 'seep' call and displays white outer tail feathers in a bounding flight. The GREY WAGTAIL *Motacilla cinerea* has regularly occurred in recent years on migration and is generally found alongside fast-flowing streams or freshwater culverts. Its slate grey back and lighter yellow breast are distinctive. The FOREST WAGTAIL *Dendronanthus indicus*

Forest Wagtail

is also regularly recorded on migration but in smaller numbers. It prefers forest habitats but may be encountered outside of these in more open ground. The black breast band and brown-and-white plumage make it a difficult bird to see on favoured leafy ground.

RICHARD'S PIPIT

Anthus novaeseelandiae 20 cm (8″) Plate 14f

A common resident, this bird occurs in open areas, preferring to feed on insects in short grass. It stands tall, with an erect posture, and has a streaked brown back and white eyebrow stripe. It will run off when disturbed and in flight exposes white outer tail feathers. A similar but more uncommon species, the RED-THROATED PIPIT *Anthus cervinus*, is a migrant of slightly smaller size. It is less upright in stance and has a high-pitched piercing 'pseeoo' call. It is mostly seen on the Tuas reclamation area, but is also recorded from Serangoon Sewerage Works and the golf course on Sentosa.

Shrikes

Of medium size with large heads, strong, hooked bills, and long tails, shrikes are commonly known as butcher birds because of their habit of catching large insect prey and storing it on a sharp thorn larder for later consumption.

BROWN SHRIKE

Lanius cristatus 20 cm (7¾″) Plate 17f

Long-tailed Shrike

A common migrant, the Brown Shrike may comprise a number of different races with variable brown plumage, a blackish mask, and white on the wings and tail. The species is easily seen perched on top of low vegetation from which it makes forays to catch larger insects with its characteristic sharp-tipped bill. It occurs throughout the island, often in vegetation along roadsides. The similar TIGER SHRIKE *Lanius tigrinus* has a heavier bill than the previous species, and distinctive barring to the flanks. It is also a migrant, but is more skulking in its habits, and is usually found in forested locations. The larger LONG-TAILED SHRIKE *Lanius schach* has a grey nape with a black burglar mask, rufous back, and black wings with a distinctive white patch. It is Singapore's only resident shrike, but occurs only in limited numbers, generally in rural wasteland areas.

Starlings and Mynas

These are medium-size birds generally of grey or black-and-white plumage. Highly gregarious, aggressive, and intelligent, they take a wide variety of foods in many different habitats. They are popular cage-birds.

PHILIPPINE GLOSSY STARLING

Aplonis panayensis 20 cm (8″) Plate 18a

A very common resident occurring in large flocks throughout both urban and rural areas, the Philippine Glossy Starling has a bright red eye and, in sunlight, an attractive glossy green plumage. The immature bird has broad, dark streaks on a white breast. The adult nests in tree holes but will also use roof eaves. Its repertoire comprises sharp, metallic whistles or mews. It feeds on berries, fruit, and insects, but is also partial to waste food. The similar PURPLE-BACKED STARLING *Sturnus sturninus* is a common migrant that mixes freely with flocks of Glossy Starlings, but is distinguished by a lighter grey head and unstreaked under-parts.

BLACK-WINGED STARLING

Sturnus melanopterus 23 cm (9″) Plate 18b

A predominately white bird with black wings and tail and a distinctive yellow eye patch, this is an introduced resident species that has a limited range and is mainly confined to St. John's Island. A flock of up to ten birds seem able to sustain themselves. Although originally comprising thirty or more individuals, numbers have reduced in recent years. It spends much time on the ground feeding on seeds and small insects, but it is also partial to fruit.

COMMON MYNA

Acridotheres tristis 25 cm (10″) Plate 18c

This species has a purplish-brown plumage and a yellow eye patch, but confusion may occur when the bird moults exposing a yellow skin head. Not quite so numerous in urban areas as its name implies, it is nevertheless still a common resident. It was first recorded in Singapore in 1936 as a result of a natural spread southwards from Malaysia.

WHITE-VENTED MYNA

Acridotheres javanicus 25 cm (10″) Plate 18d

Mainly black with a pronounced crest, yellow bill, and white wing patches seen in flight, the White-vented Myna is one of the most common and conspicuous birds in Singapore. It occurs in virtually all habitats except deep forest. Originally introduced in the 1920s, it forms large communal roosts in housing estates, much to the consternation of human residents, and nests in tree holes and roof eaves. Aggressive, assertive, and unafraid of humans, it often feeds on waste scraps, even inside houses. It issues a range of tuneless squeaks and clicking calls, and often indulges in head shaking and ruffling of feathers.

HILL MYNA
Gracula religiosa 30 cm (12″) Plate 18e

The Hill Myna has a larger-sized red bill and yellow neck lappets that distinguish it from the previous species. The bird issues a strong, penetrating 'tiong tiong' call. It is a common cage-bird that can be successfully trained to talk, but it occurs naturally in the wild in Singapore, mainly in the forests. A regular visitor to the Botanic Gardens and offshore islands, it nests in tree holes, but breeding has still to be confirmed.

Sunbirds and Spiderhunters

A beautiful family of birds, the males have a distinctive plumage of iridescent colours. They are very active at all times of the day, and use their slender, curved bills to feed on insects and nectar. Their nests comprise a delicate hanging construction of cobwebs and grasses with a side hole entrance.

BROWN-THROATED SUNBIRD
Anthreptes malacensis 14 cm (5½″) Plate 19a

Common throughout the island, the male is easily recognized by its purplish-brown throat and metallic purple shoulder patch. The female is similar to other sunbirds but differs from the Olive-backed Sunbird in having a straighter, less decurved bill and a faint yellowish spectacle around the eye. It also has no white in its plain olive tail feathers. It is a larger sunbird that frequents gardens, parks, and forest edges, and has a particular liking for coconut trees. It feeds on nectar and associated insects. It issues a series of 'chip' notes as it flits between trees in a haphazard flight pattern.

PURPLE-THROATED SUNBIRD
Nectarinia sperata 10 cm (4″) Plate 19b

Of spectacular appearance, with a deep purple throat, dark chestnut breast, and metallic green crown, this small bird is a less common resident. The female has duller yellowish-olive under-parts than other species, with a yellow streak to the centre of the belly. The narrow white tips to the black tail feathers are difficult to observe. It is usually seen in or near secondary forest areas, but has been viewed in large flocks crossing the southern waters on migration in the off-breeding period.

COPPER-THROATED SUNBIRD

Nectarinia calcostetha 14 cm (5½″) Plate 19c

This bird is similar to but larger than the previous species, and the chestnut colouring is confined to the throat and upper breast. The female has a greyish head, dark olive upper-parts, and prominent whitish tips to the tail feathers. It is a rare local resident, mainly favouring mangrove habitats and coastal scrub. It is frequently encountered along the northern coast and offshore islands in the mangroves.

OLIVE-BACKED SUNBIRD

Nectarinia jugularis 11 cm (4¼″) Plate 19d

The bright yellow breast and metallic blue throat and face of the male is distinctive. The female also has a bright yellow breast and throat, a similar olive back, but a very dark brown tail with a white tip. It is Singapore's most common sunbird, and occurs in most parks, gardens, and along the forest edge. It feeds on nectar in hibiscus and erythrina, which it extracts from the base of the flowers with its well-curved beak.

CRIMSON SUNBIRD

Aethopyga siparaja 11 cm (4¼″) Plate 19e

The male is a beautiful bird with a bright red head, iridescent purple forecrown, and purple moustache stripes seen only at close quarters. The female has a greyish-yellow wash to the breast and a yellow-olive back. A very active bird, usually inhabitating the forest edge, it issues sharp 'wheet' notes as it feeds on insects and flower nectar. As with other sunbirds, it constructs a delicate hanging nest cup of cobwebs and fine grasses.

LITTLE SPIDERHUNTER

Arachnothera longirostra 16 cm (6¼″) Plate 19f

A small, uncommon, resident bird with a white throat and a remarkably long, decurved bill used for extracting nectar from long flowers and for obtaining insects from niches in tree bark, the Little Spiderhunter is mainly confined to forested areas, and appears to have been reduced in numbers in recent years. It flies at great speed through the forest at lower levels, and issues high-pitched 'whe-che' calls.

Flowerpeckers

Small, squat birds with short tails and small bills, flowerpeckers fly quickly and erratically between the treetops where they feed on fruits, berries, and flowers.

ORANGE-BELLIED FLOWERPECKER

Dicaeum trigonostigma 9 cm (3½″) Plate 20a

The male is brilliantly coloured with a bright orange belly, back, and rump. The female is an olive green bird with a pale breast and orange rump. Immature birds have orange bills. Locally common and resident, it can often be seen feeding on the black berries of *Melastoma malabathricum* alongside the tracks of the Central Catchment Nature Reserve. Outside of this habitat it is rarely encountered, seeming to prefer being close to the forest edge.

SCARLET-BACKED FLOWERPECKER

Dicaeum cruentatum 9 cm (3½″) Plate 20b

A common resident, the male is attractively coloured with a wide crimson 'painted' stripe over the crown and down the centre of the back. The female also has a red rump and upper tail coverts but has a generally greyer plumage. It occurs more readily than previous species in parks and gardens, and can be frequently seen diving into the upper tree canopy issuing 'tik tik' calls. It feeds on mistletoe berries and also takes small insects.

Sparrows, Weavers, and Munias

Small, thick-billed seed-eaters, these birds are highly gregarious and widespread, and favour grassland habitats.

EURASIAN TREE SPARROW

Passer montanus 15 cm (6″) Plate 20c

This species is easily distinguished from the house sparrow so common in many countries, but not in Singapore, by having a brown cap and black patches on its white cheeks. It is a common, highly gregarious bird that readily occupies urban areas and is frequently in close proximity to human habitats. It has very catholic taste and feeds on scraps as well as natural grass seeds.

BAYA WEAVER

Ploceus philippinus 15 cm (6″) Plate 20d

This species is the only weaverbird in Singapore. The distinctive breeding plumage of the male shown here contrasts with the dowdy brown, indistinct, streaky plumage of the female. It constructs a most attractive hanging nest with a central bulbous chamber containing the nest cup, and a long, hanging tunnel forming the only access point. The bird nests communally, often in coconut palms, and favours open country areas with long-grass habitats. It is amusing to observe during courting, as the male, who constructs the nest, displays on top in an effort to attract female suitors. It issues a high-pitched, wheezy, rattle call.

SCALY-BREASTED MUNIA

Lonchura punctulata 11 cm (4½″) Plate 20e

This species has prominent brown-and-white scaling on the breast with a whitish vent. A common resident species, it prefers grassland habitats and moves around in flocks nibbling on grass seeds, particularly *Panicum spp*. It constructs an untidy ball nest of grass in small trees or long grasses. It may occur with other similar and related species, such as the JAVAN MUNIA *Lonchura leucogastroides*, but this species can be easily distinguished by its white belly and brown upper breast which are distinctly demarcated. This formerly introduced species often favours grassland habitats, but may also be found in long grasses along the forest edge.

CHESTNUT MUNIA

Lonchura malacca 11 cm (4½″) Plate 20f

Not such a common resident as either of the previous species, it is however more attractive, with a rich chestnut plumage except for the black head. It also uses its seed-eating bill to feed on grasses, but seems to favour more watery habitats. There is a white-headed race that occasionally occurs here, which is distinguished from the following species by having a richer chestnut plumage to the wings and breast. The WHITE-HEADED MUNIA *Lonchura maja* has similar but duller chestnut plumage and often occurs in mixed flocks with this species. It is commonly known as the 'cigar' bird, and is particularly fond of grassland areas on reclaimed land. Immature birds of most munia are extremely difficult to separate and reliance has to be placed on the close presence of mature birds, which can present a problem with mixed flocks!

Bird Conservation

HAVE you ever considered what it is like to be a wild bird? There is a general tendency to think that birds have the ultimate quality lifestyle: the ability to fly, to come and go as they please without passport, rules, or regulations, and to spend their lives eating, sleeping, and generally doing little 'work'. Some people even suggest that they can tell how happy birds are by the way they sing so joyously! However, scientific studies into bird behaviour suggest a completely different story.

Every animal in the wild survives by enacting the law of the jungle, and we only need to read the *Lord of the Flies* or *Watership Down* to know how savage that can be. A bird's lot is a hard one. Every minute of the day or night is spent searching for food, protecting itself from predators, and coping with the harsh realities of the climate. Life is a battle and survival is often brief. Changes in habitat availability and food sources can quickly spell disaster, and even extinction.

The demise of the White-bellied Woodpecker portrayed on the frontispiece of this book epitomizes the delicate balance of the natural world of Singapore. When natural habitats and ecosystems are destroyed, birds and other animals cannot simply 'move house' to another area as these are already ecologically 'full up'. The populations that the remaining areas can support are thus reduced until the gene pool is diminished to a number that becomes unsustainable. If we are serious about conserving wild birds in Singapore, some radical action needs to be taken to protect their habitats, to minimize disturbances to wild places, and to create and enhance an islandwide network of green corridor connections. Such requirements would benefit not only wildlife but also humans. The design and future management of such corridors will be critical for ensuring that wildlife biodiversity is maintained or even enhanced. Here lies an opportunity to provide a model for many urbanized cities and states throughout the world.

The future survival of wildlife is inextricably linked with our own survival because we, as animals, are also a part of the world's ecosystem.

In continuing to improve our own quality of life we need more facilities for outdoor recreation, the provision of a healthy environment that assures uncontaminated soil, clean air, and pure water, and places where we can obtain release from urban-induced stress.

In recent years a number of projects have been initiated by several of Singapore's government agencies to improve the protection and management of natural habitats and to design new open spaces. However, to date Singapore has only two protected nature reserves and one wild bird nature park. Even these have been subjected to development impacts with the recent construction of a high-rise building on the summit of Bukit Timah and a proposed golf course in the Central Catchment Nature Reserve. The twenty-eight sites listed in the Master Plan for the Conservation of Nature in Singapore are being given consideration in the planning process and some developments have been revised, withdrawn, or modified as a result. What is needed, however, is a more proactive approach with many more areas gazetted to ensure protection in perpetuity. The introduction of new statutory provisions to ensure that environment impacts are properly analysed and assessed for all major development proposals, and amended legislation to give more 'teeth' to existing wildlife laws are also urgently needed. Such legislation should meet the provisions of the recent Earth Summit in Rio to achieving economic progress in harmony with nature, and to attaining biodiversity on a sufficient scale to protect existing wildlife.

We know how valuable every square metre of Singapore's limited land mass is, but should such value only be recognized in terms of development that contributes to economic growth? What about contributions to human well-being and to wildlife needs, or to the future maintenance of life itself?

The environmental awareness that increases daily throughout the world will no doubt result in our children taking such matters more seriously than we do, because it will increasingly influence their own survival. The decisions we take today are going to affect their lifestyles, and unless we conserve the few places we have left our children will occupy a concrete jungle with perhaps only concrete birds! A greater understanding of how to protect existing ecosystems, encourage the establishment of new ones, and increase biological diversity is vital for the success of future bird conservation in Singapore. We hope that this book will make a small contribution to increasing awareness, understanding, and, above all, appreciation for and admiration of the beauty of nature and the remarkable diversity of birdlife in Singapore.

Checklist

THE following list has been compiled by the Records Committee of the Bird Group of the Nature Society Singapore (Lim Kim Seng, 1991) and includes those wild birds that have been positively identified to the satisfaction of this committee within the boundaries of the Republic of Singapore during the last seventeen years. The scientific nomenclature and order species follow King, Woodcock, and Dickinson (1975) and bracketed numbers provide a cross-reference to those used in that book. Asterisks (*) refer to species described in this book. Minor changes made in scientific taxonomy since 1975 have been included where appropriate.

Status Abbreviations

PM Passage Migrant
MB Migrant Breeder
WV Winter Visitor
NBV Non-breeding Visitor
RB Resident Breeder
R(B) Resident, Breeding not proven
A Accidental (Vagrant)
I Introduced (Escapee)
? Status Uncertain

To assist the reader in assessing the abundance status of each species the following key provides guidelines which are admitted to be rather subjective but are useful in knowing the rarity value of any particular sighting made.

● Abundant common and widespread
● Regular locally common
● Occasional uncommon
● Scarce rare

These categories relate to the likelihood of an observer sighting any particular species on a full day's outing in Singapore.

SPECIES		STATUS	ABUNDANCE

Hydrobatidae: **Storm-petrels**

No.	Species	Status	Abundance
1. (8)	Swinhoe's Storm-petrel *Oceanodroma monorhis*	PM	●

Sulidae: **Boobies**

No.	Species	Status	Abundance
2. (14)	Brown Booby *Sula leucogaster*	NBV	●

Fregatidae: **Frigatebirds**

No.	Species	Status	Abundance
3. (20)	Christmas Frigatebird *Fregata andrewsi*	NBV	●
4. (22)	Lesser Frigatebird *Fregata ariel*	NBV	●

Ardeidae: **Herons, Egrets, and Bitterns**

No.	Species	Status	Abundance
5. (24)	*Great-billed Heron *Ardea sumatrana*	RB	●
6. (25)	*Grey Heron *Ardea cinerea*	RB	●
7. (26)	*Purple Heron *Ardea purpurea*	RB WV	●
8. (27)	*Little Heron *Butorides striatus*	RB WV	●
9. (29)	*Chinese Pond-heron *Ardeola bacchus*	WV	●
10. (31)	*Cattle Egret *Bubulcus ibis*	WV	●
11. (32)	*Pacific Reef-egret *Egretta sacra*	R(B)	●
12. (33)	*Chinese Egret *Egretta eulophotes*	WV	●
13. (34)	*Great Egret *Casmerodius albus*	WV	●
14. (35)	Plumed Egret *Egretta intermedia*	WV	●
15. (36)	*Little Egret *Egretta garzetta*	WV	●
16. (37)	*Black-crowned Night-heron *Nycticorax nycticorax*	RB	●

SPECIES			STATUS	ABUNDANCE
17.	(38)	Malayan Night-heron *Gorsachius melanolophus*	WV	●
18.	(39)	*Yellow Bittern *Ixobrychus sinensis*	RB WV	●
19.	(40)	Schrenck's Bittern *Ixobrychus eurhythmus*	WV	●
20.	(41)	*Cinnamon Bittern *Ixobrychus cinnamomeus*	R(B) WV	●
21.	(42)	*Black Bittern *Ixobrychus flavicollis*	PM WV	●

Threskiornithidae: **Ibises and Spoonbills**

22.	(56)	Glossy Ibis *Plegadis falcinellus*	A	●

Anatidae: **Geese and Ducks**

23.	(64)	*Lesser Treeduck *Dendrocygna javanica*	RB	●
24.	(68)	Common Pintail *Anas acuta*	WV	●
25.	(69)	Common Teal *Anas crecca*	A	●
26.	(74)	Gadwall *Anas strepera*	A	●
27.	(76)	Eurasian Wigeon *Anas penelope*	A	●
28.	(77)	*Garganey *Anas querquedula*	WV	●
29.	(78)	Northern Shoveler *Anas clypeata*	A	●
30.	(87)	*Cotton Pygmy Goose *Nettapus coromandelianus*	NBV	●

Pandionidae: **Osprey**

31.	(95)	*Osprey *Pandion haliaetus*	NBV	●

Accipitridae: **Kites, Hawks, Vultures, and Eagles**

32.	(97)	*Black Baza *Aviceda leuphotes*	PM MV	●

SPECIES			STATUS	ABUNDANCE
33.	(98)	Crested Honey-buzzard *Pernis ptilorhyncus*	PM WV	●
34.	(99)	Bat Hawk *Macheiramphus alcinus*	NBV	●
35.	(100)	*Black-shouldered Kite *Elanus caeruleus*	RB	●
36.	(101)	*Black Kite *Milvus migrans*	WV	●
37.	(102)	*Brahminy Kite *Haliastur indus*	RB	●
38.	(103)	*White-bellied Sea-eagle *Haliaeetus leucogaster*	RB	●
39.	(107)	Grey-headed Fish-eagle *Ichthyophaga ichthyaetus*	R(B)	●
40.	(113)	Short-toed Eagle *Circaetus gallicus*	PM	●
41.	(114)	Crested Serpent-eagle *Spilornis cheela*	R(B)	●
42.	(115B)	Eastern Marsh Harrier *Circus spilonotus*	WV	●
43.	(116)	Northern Harrier *Circus cyaneus*	WV	●
44.	(119)	Pied Harrier *Circus melanoleucos*	WV	●
45.	(121)	*Japanese Sparrowhawk *Accipiter gularis*	PM WV	●
46.	(124)	Crested Goshawk *Accipiter trivirgatus*	RB NBV	●
47.	(125)	Chinese Goshawk *Accipiter soloensis*	PM	●
48.	(129)	Grey-faced Buzzard *Butastur indicus*	WV	●
49.	(130)	Common Buzzard *Buteo buteo*	WV	●
50.	(135)	Greater Spotted Eagle *Aquila clanga*	WV	●
51.	(136B)	Steppe Eagle *Aquila nipalensis*	A	●
52.	(137)	Imperial Eagle *Aquila heliaca*	A	●

SPECIES			STATUS	ABUNDANCE
53.	(140)	Booted Eagle *Hieraaetus pennatus*	A	
54.	(141)	Rufous-bellied Eagle *Hieraaetus kienerii*	WV	
55.	(142)	Changeable Hawk-eagle *Spizaetus cirrhatus*	R(B)	
56.	(144)	Blyth's Hawk-eagle *Spizaetus alboniger*	NBV	

Falconidae: **Falcons**

57.	(148)	Black-thighed Falconet *Microhierax fringillarius*	R(B)	
58.	(151)	Eurasian Kestrel *Falco tinnunculus*	WV	
59.	(158)	Peregrine Falcon *Falco peregrinus*	WV	

Phasianidae: **Quails, Partridges, and Pheasants**

60.	(164)	Blue-breasted Quail *Coturnix chinensis*	RB	
61.	(186)	Red Junglefowl *Gallus gallus*	RB	

Turnicidae: **Buttonquails**

62.	(199)	Barred Buttonquail *Turnix suscitator*	RB	

Rallidae: **Rails, Crakes, and Coots**

63.	(205)	*Slaty-breasted Rail *Rallus striatus*	RB	
64.	(206)	*Red-legged Crake *Rallina fasciata*	RB WV	
65.	(207)	Slaty-legged Crake *Rallina eurizonoides*	A	
66.	(208)	Baillon's Crake *Porzana pusilla*	WV	

SPECIES		STATUS	ABUNDANCE
67. (210)	Ruddy-breasted Crake *Porzana fusca*	RB	●
68. (213)	*White-browed Crake *Porzana cinerea*	RB	●
69. (215)	*White-breasted Waterhen *Amaurornis phoenicurus*	RB WV	●
70. (216)	*Watercock *Gallicrex cinerea*	WV	●
71. (217)	*Common Moorhen *Gallinula chloropus*	RB	●
72. (218)	*Purple Swamphen *Porphyrio porphyrio*	RB	●
73. (219)	Common Coot *Fulica atra*	A	●

Jacanidae: **Jacanas**

74. (223)	Pheasant-tailed Jacana *Hydrophasianus chirurgus*	WV	●

Rostratulidae: **Paintedsnipes**

75. (225)	Greater Paintedsnipe *Rostratula benghalensis*	RB	●

Charadriidae: **Lapwings and Plovers**

76. (228)	Grey-headed Lapwing *Hoplopterus cinereus*	A	●
77. (229)	Red-wattled Lapwing *Hoplopterus indicus*	NBV	●
78. (231)	*Grey Plover *Pluvialis squatarola*	WV PM	●
79. (232	*Pacific Golden Plover *Pluvialis fulva*	WV PM	●
80. (233)	*Common Ringed Plover *Charadrius hiaticula*	PM	●
81. (234)	*Little Ringed Plover *Charadrius dubius*	WV PM	●

SPECIES			STATUS	ABUNDANCE
82.	(235)	*Kentish Plover *Charadrius alexandrinus*	WV PM	●
83.	(236)	*Malaysian Plover *Charadrius peronii*	WV	●
84.	(238)	*Mongolian Plover *Charadrius mongolus*	WV PM	●
85.	(239)	*Greater Sand-plover *Charadrius leschenaultii*	WV PM	●
86.	(240)	Oriental Plover *Charadrius veredus*	PM	●

Scolopacidae: **Curlews, Godwits, Sandpipers, and Snipes**

SPECIES			STATUS	ABUNDANCE
87.	(241)	*Eurasian Curlew *Numenius arquata*	WV PM	●
88.	(242)	*Whimbrel *Numenius phaeopus*	WV PM	●
89.	(244)	*Eastern Curlew *Numenius madagascariensis*	PM	●
90.	(245)	*Black-tailed Godwit *Limosa limosa*	WV PM	●
91.	(246)	*Bar-tailed Godwit *Limosa lapponica*	WV PM	●
92.	(247)	Spotted Redshank *Tringa erythropus*	WV	●
93.	(248)	*Common Redshank *Tringa totanus*	WV PM	●
94.	(249)	*Marsh Sandpiper *Tringa stagnatilis*	WV PM	●
95.	(250)	*Common Greenshank *Tringa nebularia*	WV	●
96.	(251)	Nordmann's Greenshank *Tringa guttifer*	WV	●
97.	(252)	Green Sandpiper *Tringa ochropus*	WV	●
98.	(253)	*Wood Sandpiper *Tringa glareola*	WV PM	●

SPECIES			STATUS	ABUNDANCE
99.	(254)	*Terek Sandpiper *Xenus cinereus*	WV PM	●
100.	(255)	*Common Sandpiper *Actitis hypoleucos*	WV PM	●
101.	(256)	Grey-tailed Tattler *Heteroscelus brevipes*	WV PM	●
102.	(257)	Ruddy Turnstone *Arenaria interpres*	WV	●
103.	(258)	Asian Dowitcher *Limnodromus semipalmatus*	WV PM	●
104.	(261)	*Pintail Snipe *Gallinago stenura*	WV	●
105.	(262)	Swinhoe's Snipe *Gallinago megala*	WV	●
106.	(263)	*Common Snipe *Gallinago gallinago*	WV	●
107.	(265)	Eurasian Woodcock *Scolopax rusticola*	WV	●
108.	(266)	Red Knot *Calidris canutus*	WV	●
109.	(267)	Great Knot *Calidris tenuirostris*	PM	●
110.	(268)	*Rufous-necked Stint *Calidris ruficollis*	WV PM	●
111.	(270)	*Temminck's Stint *Calidris temminckii*	PM	●
112.	(271)	*Long-toed Stint *Calidris subminuta*	WV PM	●
113.	(272)	Sharp-tailed Sandpiper *Calidris acuminata*	A	●
114.	(273)	Dunlin *Calidris alpina*	PM	●
115.	(274)	*Curlew Sandpiper *Calidris ferruginea*	WV PM	●
116.	(275)	Sanderling *Calidris alba*	WV	●
117.	(276)	Spoon-billed Sandpiper *Eurynorhynchus pygmaeus*	WV	●
118.	(277)	*Broad-billed Sandpiper *Limicola falcinellus*	WV PM	●

SPECIES			STATUS	ABUNDANCE
119.	(278)	Ruff *Philomachus pugnax*	WV	●
Recurvirostridae: **Ibisbill, Stilts, and Avocets**				
120.	(280)	Black-winged Stilt *Himantopus himantopus*	WV PM	●
Burhinidae: **Thick-knees**				
121.	(285)	Beach Thick-knee *Esacus magnirostris*	R(B)	●
Glareolidae: **Pratincoles**				
122.	(286)	Oriental Pratincole *Glareola maldivarum*	PM	●
123.	(287)	Small Pratincole *Glareola lactea*	A	●
Stercorariidae: **Jaegers and Skuas**				
124.	(288-1)	Parasitic Jaeger *Stercorarius parasiticus*	A	●
Laridae: **Gulls and Terns**				
125.	(290)	Common Black-headed Gull *Larus ridibundus*	WV	●
126.	(291)	Brown-headed Gull *Larus brunnicephalus*	WV	●
127.	(298)	*Whiskered Tern *Chlidonias hybrida*	A	●
128.	(299)	*White-winged Tern *Chlidonias leucopterus*	WV PM	●
129.	(300)	Gull-billed Tern *Gelochelidon nilotica*	WV PM	●
130.	(301)	Caspian Tern *Hydroprogne caspia*	WV	●
131.	(303)	Common Tern *Sterna hirundo*	WV PM?	●

SPECIES			STATUS	ABUNDANCE
132.	(305)	*Black-naped Tern *Sterna sumatrana*	RB	●
133.	(307)	Bridled Tern *Sterna anaethetus*	WV	●
134.	(309)	*Little Tern *Sterna albifrons*	RB WV	●
135.	(310)	*Great Crested Tern *Sterna bergii*	WV	●
136.	(310)	*Lesser Crested Tern *Sterna bengalensis*	WV	●

Columbidae: **Pigeons and Doves**

137.	(321)	*Thick-billed Pigeon *Treron curvirostra*	R(B) NBV	●
138.	(323)	Cinnamon-headed Pigeon *Treron fulvicollis*	NBV	●
139.	(324)	Little Green Pigeon *Treron olax*	R(B) NBV	●
140.	(325)	*Pink-necked Pigeon *Treron vernans*	RB	●
141.	(329)	Jambu Fruit-Dove *Ptilinopus jambu*	NBV	●
142.	(331)	Pied Imperial Pigeon *Ducula bicolor*	NBV	●
143.	(334)	Feral Pigeon *Columba livia*	I	●
144.	(342)	*Red Turtle-dove *Streptopelia tranquebarica*	I	●
145.	(343)	*Spotted Dove *Streptopelia chinensis*	RB	●
146.	(344)	*Peaceful Dove *Geopelia striata*	RB	●
147.	(345)	*Green-winged Pigeon *Chalcophaps indica*	R(B)	●

Psittacidae: **Parrots**

148.	(348)	Rose-ringed Parakeet *Psittacula krameri*	I	●

SPECIES			STATUS	ABUNDANCE
149.	(349)	Red-breasted Parakeet *Psittacula alexandri*	I	●
150.	(350)	*Long-tailed Parakeet *Psittacula longicauda*	RB	●
151.	(353)	Blue-rumped Parrot *Psittinus cyanurus*	R(B)	●
152.	(355)	Blue-crowned Hanging Parrot *Loriculus galgulus*	R(B)	●
153.		Lesser Sulphur-crested Cockatoo *Cacatua sulphurea*	I	●
154.		Goffin's Cockatoo *Cacatua goffini*	I	●

Cuculidae: **Cuckoos**

155.	(356)	Chestnut-winged Cuckoo *Clamator coromandus*	WV PM	●
156.	(358)	Large Hawk-cuckoo *Cuculus sparverioides*	WV	●
157.	(361)	Hodgson's Hawk-cuckoo *Cuculus fugax*	WV	●
158.	(362)	*Indian Cuckoo *Cuculus micropterus*	WV	●
159.	(366)	*Banded Bay Cuckoo *Cacomantis sonneratii*	RB	●
160.	(367)	*Plaintive Cuckoo *Cacomantis merulinus*	RB	●
161.	(368)	Indonesian Cuckoo *Cacomantis sepulcralis*	R(B) NBV?	●
162.	(370)	Violet Cuckoo *Chrysococcyx xanthorhynchus*	R(B) WV	●
163.	(371)	Horsfield's Bronze Cuckoo *Chrysococcyx basalis*	WV	●
164.	(372)	Malayan Bronze Cuckoo *Chrysococcyx minutillus*	RB	●
165.	(373)	Drongo Cuckoo *Surniculus lugubris*	R(B) WV	●

SPECIES			STATUS	ABUNDANCE
166.	(374)	*Common Koel *Eudynamys scolopacea*	WV	●
167.	(376)	*Chestnut-bellied Malkoha *Phaenicophaeus sumatranus*	RB	●
168.	(383)	*Greater Coucal *Centropus sinensis*	RB	●
169.	(384)	*Lesser Coucal *Centropus bengalensis*	RB	●

Tytonidae: **Barn and Bay Owls**

170.	(385)	*Barn Owl *Tyto alba*	RB	●

Strigidae: **Owls**

171.	(391)	Oriental Scops-owl *Otus sunia*	WV	●
172.	(392)	*Collared Scops-owl *Otus bakkamoena*	RB	●
173.	(399)	Buffy Fish-owl *Ketupa ketupu*	R(B)	●
174.	(403)	*Brown Hawk-owl *Ninox scutulata*	RB WV	●
175.	(405)	*Spotted Wood-owl *Strix seloputo*	RB	●
176.	(410)	Short-eared Owl *Asio flammeus*	WV	●

Caprimulgidae: **Nightjars**

177.	(416)	*Malaysian Eared Nightjar *Eurostopodus temminckii*	R(B)	●
178.	(418)	Grey Nightjar *Caprimulgus indicus*	PM WV?	●
179.	(419)	*Large-tailed Nightjar *Caprimulgus macrurus*	RB	●
180.	(421)	Savanna Nightjar *Caprimulgus affinis*	RB	●

SPECIES			STATUS	ABUNDANCE
***Apodidae*: Swifts**				
181.	(423)	*Edible-nest Swiftlet *Aerodramus fuciphagus*	RB	●
182.	(424)	*Black-nest Swiftlet *Aerodramus maximus*	RB	●
183.	(426)	White-bellied Swiftlet *Collocalia esculenta*	A	●
184.	(428)	White-vented Needletail *Hirundapus cochinchinensis*	PM	●
185.	(429)	Brown Needletail *Hirundapus giganteus*	PM WV	●
186.	(432)	*Fork-tailed Swift *Apus pacificus*	PM	●
187.	(433)	*House Swift *Apus affinis*	RB	●
188.	(434)	*Asian Palmswift *Cypsiurus balasiensis*	RB	●
***Hemiprocnidae*: Treeswifts**				
189.	(436)	*Grey-rumped Treeswift *Hemiprocne longipennis*	RB	●
190.	(437)	Whiskered Treeswift *Hemiprocne comata*	NBV A	●
***Alcedinidae*: Kingfishers**				
191.	(448)	*Common Kingfisher *Alcedo atthis*	WV	●
192.	(449)	*Blue-eared Kingfisher *Alcedo meninting*	R(B)	●
193.	(451)	*Oriental Dwarf Kingfisher *Ceyx erithacus*	WV	●
194.	(454)	*Stork-billed Kingfisher *Halcyon capensis*	R(B)	●
195.	(456)	*Ruddy Kingfisher *Halcyon coromanda*	RB WV	●

SPECIES			STATUS	ABUNDANCE
196.	(457)	*White-throated Kingfisher *Halcyon smyrnensis*	RB	●
197.	(458)	*Black-capped Kingfisher *Halcyon pileata*	WV	●
198.	(459)	*Collared Kingfisher *Halcyon chloris*	RB	●
Meropidae: **Bee-eaters**				
199.	(462)	*Blue-tailed Bee-eater *Merops superciliosus*	WV	●
200.	(464)	*Blue-throated Bee-eater *Merops viridis*	MB WV	●
Coraciidae: **Rollers**				
201.	(468)	*Dollarbird *Eurystomus orientalis*	RB WV	●
Capitonidae: **Barbets**				
202.	(489)	*Red-crowned Barbet *Megalaima rafflesii*	RB	●
203.	(497)	*Coppersmith Barbet *Megalaima haemacephala*	RB	●
Picidae: **Woodpeckers**				
204.	(505)	*Rufous Woodpecker *Celeus brachyurus*	RB	●
205.	(506)	*Laced Woodpecker *Picus vittatus*	RB	●
206.	(516)	*Banded Woodpecker *Picus miniaceus*	RB	●
207.	(519)	*Common Goldenback *Dinopium javanense*	RB	●
208.	(526)	Great Slaty Woodpecker *Mulleripicus pulverulentus*	NBV	●

SPECIES			STATUS	ABUNDANCE
209.	(527)	*White-bellied Woodpecker *Dryocopus javensis*	R(B)	●
210.	(536)	*Brown-capped Woodpecker *Picoides moluccensis*	RB	●

Pittidae: **Pittas**

211.	(554)	*Blue-winged Pitta *Pitta moluccensis*	WV PM	●
212.	(554)	*Mangrove Pitta *Pitta megarhyncha*	RB	●
213.	(556)	*Hooded Pitta *Pitta sordida*	WV PM	●

Hirundinidae: **Swallows**

214.	(570)	Sand Martin *Riparia riparia*	WV PM	●
215.	(572)	*Barn Swallow *Hirundo rustica*	WV PM	●
216.	(573)	*Pacific Swallow *Hirundo tahitica*	RB	●
217.	(575)	Red-rumped Swallow *Hirundo daurica*	PM	●
218.	(577)	Asian House-martin *Delichon dasypus*	WV PM	●

Camperhagidae: **Cuckoo-shrikes and Minivets**

219.	(587)	Lesser Cuckoo-shrike *Coracina fimbriata*	R(B)? A?	●
220.	(589)	*Pied Triller *Lalage nigra*	RB	●
221.	(590)	*Ashy Minivet *Pericrocotus divaricatus*	WV	●
222.	(598)	*Scarlet Minivet *Pericrocotus flammeus*	R(B)	●

		SPECIES	STATUS	ABUNDANCE
		Chloropseidae: **Ioras and Leafbirds**		
223.	(600)	*Common Iora *Aegithina tiphia (singapurensis)*	RB	●
224.	(602)	*Lesser Green Leafbird *Chloropsis cyanopogon*	R(B)	●
225.	(603)	*Greater Green Leafbird *Chloropsis sonneratii*	R(B)? NBV?	●
226.	(605)	*Blue-winged Leafbird *Chloropsis cochinchinensis*	R(B)	●
		Pycnonotidae: **Bulbuls**		
227.	(609)	*Straw-headed Bulbul *Pycnonotus zeylanicus*	RB	●
228.	(612)	Black-headed Bulbul *Pycnonotus atriceps*	R(B)	●
229.	(616)	Red-whiskered Bulbul *Pycnonotus jocosus*	I	●
230.	(624)	*Yellow-vented Bulbul *Pycnonotus goiavier*	RB	●
231.	(625)	*Olive-winged Bulbul *Pycnonotus plumosus*	RB	●
232.	(627)	*Cream-vented Bulbul *Pycnonotus simplex*	R(B)	●
233.	(628)	*Red-eyed Bulbul *Pycnonotus brunneus*	R(B)	●
234.	(639)	Buff-vented Bulbul *Hypsipetes charlottae*	A? R(B)?	●
235.	(641)	Streaked Bulbul *Hypsipetes malaccensis*	A?	●
236.	(642)	Ashy Bulbul *Hyposipetes flavala*	NBV	●
		Dicruridae: **Drongos**		
237.	(646)	*Black Drongo *Dicrurus macrocercus*	WV	●
238.	(647)	Ashy Drongo *Dicrurus leucophaeus*	WV	●

SPECIES			STATUS	ABUNDANCE
239.	(648)	*Crow-billed Drongo *Dicrurus annectans*	WV PM	●
240.	(652)	*Greater Racket-tailed Drongo *Dicrurus paradiseus*	RB	●

Oriolidae: **Old World Orioles and Fairy-bluebirds**

241.	(654)	*Black-naped Oriole *Oriolus chinensis*	RB WV	●
242.	(660)	*Asian Fairy-bluebird *Irena puella*	RB	●

Corvidae: **Jays, Magpies, and Crows**

243.	(677)	*House Crow *Corvus splendens*	I	●
244.	(681)	*Large-billed Crow *Corvus macrorhynchos*	RB	●

Timaliidae: **Babblers**

245.	(715)	*Short-tailed Babbler *Trichastoma malaccense*	R(B)	●
246.	(716)	White-chested Babbler *Trichastoma rostratum*	R(B)	●
247.	(719)	*Abbott's Babbler *Trichastoma abbotti*	RB	●
248.	(720)	Moustached Babbler *Malacopteron magnirostre*	RB	●
249.	(759)	*Chestnut-winged Babbler *Stachyris erythroptera*	RB	●
250.	(760)	*Striped Tit-babbler *Macronous gularis*	RB	●
251.	(792)	Hwamei *Garrulax canorus*	I	●

Turdidae: **Thrushes**

252.	(874)	Siberian Blue Robin *Erithacus cyane*	WV PM	●

SPECIES			STATUS	ABUNDANCE
253.	(879)	*Magpie Robin *Copsychus saularis*	RB	●
254.	(880)	*White-rumped Shama *Copsychus malabaricus*	R(B)	●
255.	(900)	Stonechat *Saxicola torquata*	WV	●
256.	(907)	White-throated Rock-thrush *Monticola gularis*	WV PM	●
257.	(913)	Orange-headed Thrush *Zoothera citrina*	WV	●
258.	(914)	Siberian Thrush *Zoothera sibirica*	WV PM	●
259.	(930)	Eye-browed Thrush *Turdus obscurus*	PM	●

Sylvidae: **Old World Warblers**

SPECIES			STATUS	ABUNDANCE
260.	(933)	*Flyeater *Gerygone sulphurea*	RB	●
261.	(950)	Inornate Warbler *Phylloscopus inornatus*	WV	●
262.	(953)	*Arctic Warbler *Phylloscopus borealis*	WV PM	●
263.	(957)	Eastern Crowned Warbler *Phylloscopus coronatus*	WV PM	●
264.	(965)	*Oriental Reed-warbler *Acrocephalus orientalis*	WV PM	●
265.	(966)	*Black-browed Reed-warbler *Acrocephalus bistrigiceps*	WV	●
266.	(970)	*Pallas's Warbler *Locustella certhiola*	WV PM	●
267.	(972)	*Lanceolated Warbler *Locustella lanceolata*	WV PM	●
268.	(975)	*Common Tailorbird *Orthotomus sutorius*	RB	●
269.	(976)	*Dark-necked Tailorbird *Orthotomus atrogularis*	RB	●
270.	(977)	*Ashy Tailorbird *Orthotomus sepium*	RB	●

SPECIES			STATUS	ABUNDANCE
271.	(978)	*Rufous-tailed Tailorbird *Orthotomus sericeus*	RB	●
272.	(983)	*Yellow-bellied Prinia *Prinia flaviventris*	RB	●
273.	(987)	*Zitting Cisticola *Cisticola juncidis*	RB	●

Muscicapidae: **Old World Flycatchers**

SPECIES			STATUS	ABUNDANCE
274.	(1005)	Brown-chested Flycatcher *Rhinomyias brunneata*	WV PM	●
275.	(1007)	Dark-sided Flycatcher *Muscicapa sibirica*	WV PM	●
276.	(1009)	*Asian Brown Flycatcher *Muscicapa dauurica*	WV PM	●
277.	(1012)	*Ferruginous Flycatcher *Muscicapa ferruginea*	WV	●
278.	(1014)	Yellow-rumped Flycatcher *Ficedula zanthopygia*	WV PM	●
279.	(1016)	Mugimaki Flycatcher *Ficedula mugimaki*	PM	●
280.	(1028)	Blue-and-white Flycatcher *Cyanoptila cyanomelana*	WV	●
281.	(1044)	*Mangrove Blue Flycatcher *Cyornis rufigastra*	RB	●
282.	(1051)	*Pied Fantail *Rhipidura javanica*	RB	●
283.	(1052)	Black-naped Monarch *Hypothymis azurea*	R(B)	●
284.	(1055)	*Japanese Paradise Flycatcher *Terpsiphone atrocaudata*	PM	●
285.	(1056)	*Asian Paradise Flycatcher *Terpsiphone paradisi*	WV PM	●

SPECIES	STATUS	ABUNDANCE
Rachycephalidae: **Whistlers**		
286. (1057) *Mangrove Whistler *Pachycephala cinerea*	RB	●
Motacillidae: **Wagtails and Pipits**		
287. (1061) White Wagtail *Motacilla alba*	WV PM	●
288. (1062) *Grey Wagtail *Motacilla cinerea*	WV PM	●
289. (1063) *Yellow Wagtail *Motacilla flava*	WV PM	●
290. (1064) Citrine Wagtail *Motacilla citreola*	A	●
291. (1065) *Forest Wagtail *Dendronanthus indicus*	WV PM	●
292. (1067) *Richard's Pipit *Anthus novaeseelandiae*	RB	●
293. (1069) *Red-throated Pipit *Anthus cervinus*	PM	●
Laniidae: **Shrikes**		
294. (1075) *Brown Shrike *Lanius cristatus*	WV PM	●
295. (1077) *Tiger Shrike *Lanius tigrinus*	WV PM	●
296. (1080) *Long-tailed Shrike *Lanius schach*	RB	●
Sturnidae: **Starlings and Mynas**		
297. (1082) *Philippine Glossy Starling *Aplonis panayensis*	RB	●
298. (1086) White-shouldered Starling *Sturnus sinensis*	WV	●
299. (1087) *Purple-backed Starling *Sturnus sturninus*	WV	●

SPECIES			STATUS	ABUNDANCE
300.		*Black-winged Starling *Sturnus melanopterus*	I	●
301.		Chestnut-cheeked Starling *Sturnus philippensis*	A	●
302.	(1093)	*Common Myna *Acridotheres tristis*	RB	●
303.	(1095)	*White-vented Myna *Acridotheres javanicus*	I	●
304.	(1097)	Crested Myna *Acridotheres cristatellus*	I	●
305.	(1098)	*Hill Myna *Gracula religiosa*	R(B)	●

Nectariniidae: **Sunbirds and Spiderhunters**

SPECIES			STATUS	ABUNDANCE
306.	(1100)	Plain Sunbird *Anthreptes simplex*	A? R(B)?	●
307.	(1101)	*Brown-throated Sunbird *Anthreptes malacensis*	RB	●
308.	(1106)	*Purple-throated Sunbird *Nectarinia sperata*	RB	●
309.	(1107)	*Copper-throated Sunbird *Nectarinia calcostetha*	RB	●
310.	(1108)	*Olive-backed Sunbird *Nectarinia jugularis*	RB	●
311.	(1114)	*Crimson Sunbird *Aethopyga siparaja*	RB	●
312.	(1117)	*Little Spiderhunter *Arachnothera longirostra*	RB	●
313.	(1118)	Thick-billed Spiderhunter *Arachnothera crassirostris*	NBV	●

Dicaeidae: **Flowerpeckers**

SPECIES			STATUS	ABUNDANCE
314.	(1127)	Thick-billed Flowerpecker *Dicaeum agile*	A?	●
315.	(1129)	Yellow-vented Flowerpecker *Dicaeum chrysorrheum*	R(B)?	●

SPECIES	STATUS	ABUNDANCE
316. (1131) *Orange-bellied Flowerpecker *Diaceum trigonostigma*	RB	●
317. (1134) *Scarlet-backed Flowerpecker *Diaceum cruentatum*	RB	●

Ploceidae: **Sparrows and Weavers**

SPECIES	STATUS	ABUNDANCE
318. (1140) *Eurasian Tree Sparrow *Passer montanus*	RB(I)?	●
319. (1144) *Baya Weaver *Ploceus philippinus*	RB	●

Estrildidae: **Munias**

SPECIES	STATUS	ABUNDANCE
320. (1150) Java Sparrow *Padda oryzivora*	I	●
321. (1151) White-rumped Munia *Lonchura striata*	R(B)	●
322. (1153) *Javan Munia *Lonchura leucogastroides*	I	●
323. (1154) *Scaly-breasted Munia *Lonchura punctulata*	RB	●
324. (1155) *Chestnut Munia *Lonchura malacca*	RB	●
325. (1156) *White-headed Munia *Lonchura maja*	RB	●

Emberizidae: **Finches and Buntings**

SPECIES	STATUS	ABUNDANCE
326. (1185) Yellow-breasted Bunting *Emberiza aureola*	WV	●

Select Bibliography

Briffett, C., *A Guide to the Common Birds of Singapore*, Singapore Science Centre, Singapore, 1984; revised edition 1992.

Bucknill, A. S. and Chasen, F. N., *Birds of Singapore and South-East Asia*, Government Printing Office, Singapore, 1927; reprinted Graham Brash, Singapore, 1990.

Chew Yen Fook, *Wild Birds of Singapore: The Migratory Season*, Koh Po Lin Associate, Singapore, 1989.

———, *Birds in Our Midst*, BP, Singapore, 1991.

Choo-Toh Get Ten et al., *A Guide to the Bukit Timah Nature Reserve*, Singapore Science Centre, Singapore, 1985.

Delacour, J., *Birds of Malaya*, Macmillan & Co., 1947.

Glenister, A. G., *The Birds of the Malay Peninsula, Singapore and Penang*, Oxford University Press, Kuala Lumpur, 1951.

Hails, C. and Jarvis, F., *Birds of Singapore*, Times Editions, Singapore, 1987.

King, B., Woodcock, M., and Dickinson, E. C., *A Field Guide to the Birds of South-east Asia*, Collins, London, 1975.

Lim Kim Seng, *Vanishing Birds of Singapore*, Nature Society Singapore, 1992.

Madoc, G. C., *An Introduction to Malayan Birds*, Malayan Nature Society, Kuala Lumpur, 1956.

Medway, Lord and Wells, D. R., *The Birds of the Malay Peninsula*, H. F. and G. Whitherby Ltd. and Penerbit Universiti Malaya, Kuala Lumpur, 1976.

Strange, M. and Jeyarajasingam, A., *A Photographic Guide to the Birds of Peninsular Malaysia and Singapore*, Sun Tree Publishing, Singapore, 1993.

Tweedie, M. W. F., *Common Birds of the Malay Peninsula*, 2nd edn., Longmans, Singapore, 1970.

Index

References in bold refer to main descriptions of bird species; those in italics refer to Colour Plate numbers.